Earth Resistance for Archaeologists

Geophysical Methods for Archaeology

Series Editors:
Lawrence B. Conyers, University of Denver
Kenneth L. Kvamme, University of Arkansas

The Geophysical Methods for Archaeology series aims to introduce field archaeologists and their students to the theory and methods associated with near-surface geophysical data collection techniques. Each book in this series will describe one of these commonly used non-invasive techniques, its applications, and its importance to archaeological practice for the non-specialist.

Volume 1: *Ground-Penetrating Radar for Archaeology*, Lawrence B. Conyers

Volume 2: *Magnetometry for Archaeologists*, Arnold Aspinall, Chris Gaffney, and Armin Schmidt

Volume 3: *Earth Resistance for Archaeologists*, Armin Schmidt

Earth Resistance for Archaeologists

Armin Schmidt

ALTAMIRA
PRESS
A Division of Rowman & Littlefield Publishers, Inc.
Lanham • New York • Toronto • Plymouth, UK

Published by AltaMira Press
A division of Rowman & Littlefield Publishers, Inc.
A wholly owned subsidiary of The Rowman & Littlefield Publishing Group, Inc.
4501 Forbes Boulevard, Suite 200, Lanham, Maryland 20706
www.rowman.com

10 Thornbury Road, Plymouth PL6 7PP, United Kingdom

British Library Cataloguing in Publication Information Available

Library of Congress Cataloging-in-Publication Data

Schmidt, Armin, 1961–
Earth resistance for archaeologists / Armin Schmidt.
pages cm. — (Geophysical methods for archaeology ; volume 3)
Includes bibliographical references and index.
ISBN 978-0-7591-1204-9 (cloth : alk. paper) — ISBN 978-0-7591-2293-2 (electronic) 1. Earth resistance in archaeology. I. Title.
CC79.G46S35 2013
930.1—dc23 2012043069

∞™ The paper used in this publication meets the minimum requirements of American National Standard for Information Sciences—Permanence of Paper for Printed Library Materials, ANSI/NISO Z39.48-1992.

Printed in the United States of America

To Doris

CONTENTS

LIST OF FIGURES

LIST OF TABLES

ACKNOWLEDGMENTS

First of all, I would like to thank Prof. Arnold Aspinall, Dr. Chris Gaffney, and Rob Fry for reading the entire manuscript and giving excellent comments. It was Arnold who converted me from a pure magnetist to a multimethod enthusiast—many thanks! I also wish to thank my editor, Wendi Schnaufer, for her patience, encouragement, and interest in this book. The book is also dedicated to my students, whose enthusiasm for archaeological geophysics often matched my own. Thanks also to Prof. Robin Coningham, Dr. Jan Harding, and Dr. Rick Jones for leading the archaeology teams on the case studies presented here and for their permission to use the relevant data. Thanks also to Dr. Roger Walker from Geoscan Research for providing access to the MSP40, for permission to use and reproduce data, and for insightful general support. Stuart Fox, thanks a lot, for fabulous technical support, including on fieldwork spent more often than not in pouring rain, and to Jane Hammond for rallying the postgraduate students. Thanks to my father, Siegfried, for constantly reminding me that the book isn't finished yet, and to my wife for unwavering encouragement despite a complete lack of interest in archaeological geophysics. Large parts of the book were written during my time at Queen's College, University of Melbourne—many thanks for the excellent research environment and friendship! The book was finished at my parents-in-law's house; thanks to Albert and Mathilde for their hospitality. Last but not least, thanks to the University of Bradford, Department of Archaeological Sciences, for being an invigorating place for seventeen years of my academic career, and for providing the resources to collect and process the geophysical data.

INTRODUCTION

"Trailing wires"—that's probably the first thing that comes to mind when thinking of electrical methods in archaeology. But there is much more to it. While magnetometer surveys are the "workhorse of archaeological geophysics" (Clark 1996; Gaffney and Gater 2003; Aspinall et al. 2008), and ground penetrating radar (GPR) has become the favorite technique (Conyers 2004) of many archaeological geophysicists, earth resistance methods are key to detecting many other archaeological remains, yet they have not been given nearly enough attention.

This book will redress the balance, giving archaeologists the know-how required to exploit the potential of earth resistance methods. A wide variety of possible uses are explored, including cases in which earth resistance surveys succeeded in mapping buried archaeological remains that magnetometer surveys were unable to detect, or in which they provided a very different view of the ground than GPR surveys could. The examples include prehistoric henges, Buddha's palace at Tilaurakot, Roman forts, Scottish brochs, and many other sites where new archaeological insights were gained from carefully interpreted earth resistance data. The archaeological features that can be detected through earth resistance methods are varied, ranging from ditches, pits, and grave cuts to stone and brick foundations, and even including whole landscapes. While area surveys were traditionally the most common earth resistance method, depth profiling and vertical imaging have become well-developed tools that allow electrical depth investigations in three dimensions. Both measurement techniques will be described in detail.

The challenges faced when using earth resistance surveys are best highlighted in comparison to magnetometer investigations of archaeological remains.

- Magnetic properties of archaeological features are nearly independent of current environmental factors, and results of a magnetometer survey will look the same, whatever the weather. It is quite the opposite for electrical investigations. Earth resistance anomalies are created by moisture variations in the ground and can even depend on weather events that happened several weeks prior to a survey, similar to cropmarks in aerial archaeology. It is therefore crucial to understand how these anomalies are created so that realistic archaeological interpretations can be obtained.
- As with most other geophysical techniques, visualized data from earth resistance surveys are not direct images of the underlying archaeological features. Instead, they are created as a result of the measurement process, which needs to be understood to derive the outlines of the buried remains.
- Magnetic fields of archaeological features exist whether measured or not, and in its simplest form there is not much more to a magnetometer survey than to "switch on the instrument and go." Earth resistance measurements are "active" (i.e., create their own electrical fields), and there are many survey parameters to adjust. For example, one needs to decide on the best electrode configuration for a site or the earth resistance meter has to be adjusted to allow measurements even in difficult conditions.

The knowledge required to interpret earth resistance anomalies and to select instrument configurations will be conveyed in this book. It provides a foundation from which to explore these issues without requiring a degree in physics or mathematics. A balance between the underlying principles and practical examples has been chosen in order to explain the relationship between parameters with as few mathematical expressions as possible. Instead, easy-to-understand analogies are used to illuminate important principles.

Chapter 1 introduces the most relevant electrical quantities using various examples. Following the one-dimensional case of a wire, equipotential surfaces and current lines are presented as an aid to visualize current flow in two and three dimensions. Ground resistivity and its dependence on environmental factors (e.g., moisture content, ion concentrations, evapotranspiration, grain size) are discussed, followed by an investigation of how archaeological features influence the soil's electrical resistivity.

In chapter 2 the earth resistance of homogeneous ground is examined, including its dependence on electrode arrangements and ground resistivity. This makes it easier to appreciate the variability caused by buried archaeological features, discussed subsequently.

Chapter 3 investigates the conditions of a heterogeneous ground. It explains how measurements can be characterized by using "apparent resistivity" and examines the refraction of current lines at interfaces between layers. This leads to a discussion of two important cases. The "buried sphere" is a model for a localized object, such as a stone foundation, and the "layered earth" approximates laterally extended features, such as floors. In addition to horizontal mapping of these features, it will be explained how measurements can be made with expanding electrode arrays to provide information about their depth.

Chapter 4 looks at resistivity depth imaging and the visualization of resulting data as pseudosections, as well as further processing through inversion. The discussion of three-dimensional inversion schemes leads to a detailed examination of resistivity tomography, where advanced processing techniques are required to analyze data that probe the ground with different electrode configurations.

Chapter 5 combines all the theoretical and practical information conveyed so far and discusses instrument design and field practice. Criteria for the best choice of electrode arrays are developed, and demands on earth resistance meters are explained. The earlier examination of electrode arrays is used to evaluate, for example, how far remote electrodes should be away from the survey area in a twin-probe arrangement or which variations are encountered when moving electrode arrays across a site.

Chapter 6 introduces the main data treatment techniques that can be used to improve earth resistance data so that survey defects or the effects of operator errors are minimized. Subsequent data processing with high- and low-pass filtering is discussed, carefully evaluating the respective merits and pitfalls.

In chapter 7 four case studies are presented, highlighting different aspects of earth resistance surveys and describing the processing steps that proved most useful. The interpretation of these survey results is based on an understanding of the data as developed in the preceding chapters.

An outlook with concluding considerations is presented in chapter 8, followed by a bibliography.

The material is presented in a sequence that progresses from simple cases, such as the homogeneous ground, to the complexity of buried archaeological features. At the same time, the chapters are written in a way that should allow readers to dip into those parts that are most relevant to them without having to read all the preceding material. All parts of this book have been "tested" on numerous cohorts of postgraduate students.

There are several geophysical techniques that involve the flow of electrical currents in the ground, and this book concentrates on the most relevant ones for archaeological prospecting, earth resistance surveys, and resistivity imaging. Induced polarization and self-potential methods are very rarely used in archaeology and are therefore only briefly mentioned. The induction effects of electromagnetic (EM) methods that are created by alternating currents (AC) are based on very different physical principles from those discussed here and are therefore not included.

The archiving and reporting of data, including metadata creation, has already been covered in detail elsewhere (Schmidt and Ernenwein 2011). Earth resistance area surveys have been used in archaeology for a considerable time, and their history has already been well charted (Scollar et al. 1990; Clark 1996; Hesse 2000).

Is this book also suitable for nonarchaeologists? Yes, most definitely. While it mainly includes archaeological case studies, the underlying principles are exactly the same for environmental applications. Whether investigating a silted monastic fishpond or a landfill site with resistivity imaging, the principles are the same; only the subsequent data processing steps may differ slightly. Environmental scientists who want to learn about electrical methods of investigation will therefore find much relevant information in this book.[1]

I hope that this book will serve as a helpful resource for archaeologists, planning officers, environmental scientists, researchers, and students who want to explore, and exploit, the great potential of earth resistance methods in archaeology and for environmental applications.

1. Although the text mostly refers to buried features as being "archaeological," these could easily have been called "archaeological and environmental," but that would have made the text too cumbersome to read.

CHAPTER ONE
ELECTRICAL CURRENTS IN THE GROUND

The basic idea behind earth resistance surveys is that buried archaeological features alter the flow of an electrical current that is injected into the ground through electrodes and that the effect of these changed currents can be measured at the surface. To interpret the resulting data as archaeological features requires an understanding of the relationship between what is injected (electrical current), what is measured (electrical potential), and what causes these changes (electrical resistivity). These quantities are introduced in this chapter using a conducting wire as an example. The findings are then generalized to describe the situation in the three-dimensional ground, as it is encountered in archaeological and environmental applications. A less intuitive approach that would start from a general theoretical description and would then move to the specific example of a wire is left to textbooks of physics and geophysics (Keller and Frischknecht 1966; Jackson 1975; Kaufman and Anderson 2010).

1.1 Electrical quantities

1.1.1 Basic concepts of electricity

The necessary electrical quantities can be introduced in analogy to the flow of water in a hosepipe (figure 1.1 and table 1.1), bearing in mind that such analogies are for illustrative purposes only.

The flow of water is carried by water molecules (H_2O), which can be characterized by their mass. In the electrical case the flow is carried

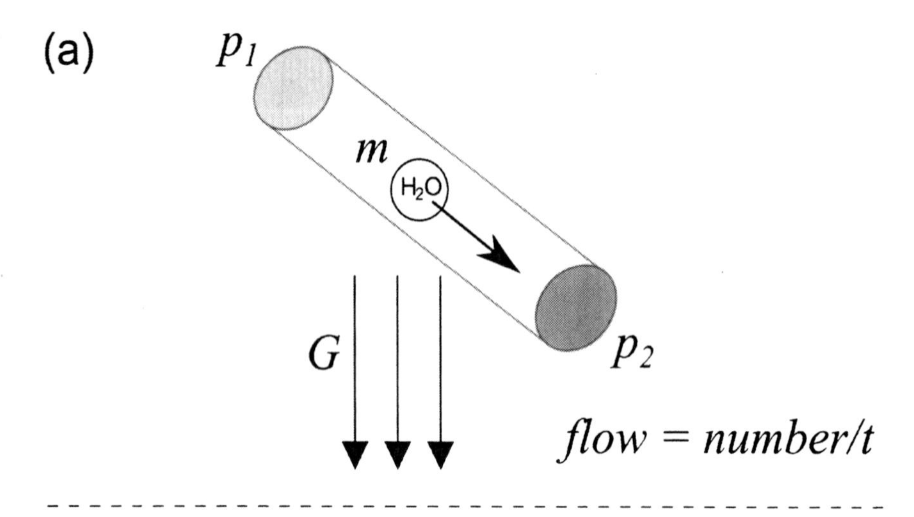

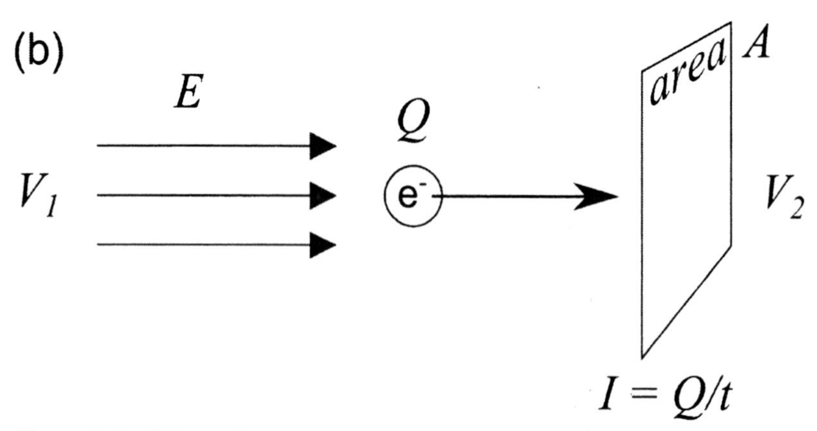

Figure 1.1. Water flowing through a tube is an analogy for electrical quantities: (a) water molecules moving through a tube in the gravitational field; (b) electrical charges moving due to an applied electrical field.

by charged particles. The property of "electrical charge" determines their movement and is often referred to with the symbol Q. It is measured in units of COULOMB (abbreviated as C). The carriers of electrical flow can be either electrons, which always have a negative charge, or ions, which can be positively or negatively charged. Since the charge of ions is created by losing or gaining electrons, they are always in multiples of an electron's charge of 1.6×10^{-9} C. The two possible polarities of charged particles allow electrical transport in two different directions, whereas the always positive mass of a molecule leaves only one possible direction for the flow of water in a downpipe.

The easiest way to describe water flow is to count the number n of molecules that pass through the end of the hosepipe during a time interval of duration t. The flow rate $f = n / t$ (i.e., number per seconds) can then be used to describe how fast the water flows. In the electrical case, the number of carriers is evaluated by measuring their electrical charge (which is always in multiples of an electron's charge, see above), and the "electrical current" is defined as $I = Q / t$ (unit AMPERE, A) to describe the flow of the charged particles. If rapid variations of the current are to be recorded, it is necessary to take measurements over very short time spans. This is expressed in the differential definition for the electrical current as $I = dQ / dt$.

The water may be driven through the hosepipe by a pressure head, which exists if there is different pressure between the two ends (p_1 and p_2). In the electrical case this is described by the "electrical potential" (V), measured in VOLTS (V).[1] Applying different potentials to the ends of a wire—for example, by using the two poles of a battery—drives a current through it. An alternative description of how the flow is created uses fields of force. The gravitational field, for example, is produced by the mass of the earth, and any other particle in the earth's gravitational field experiences an attraction in this field. Hence, in an unconnected hosepipe the water runs downhill toward the center of the earth and produces a flow of water. When measuring the pressure at both ends of the hosepipe, they would be found to be different. In a similar way, an "electrical field" (E) is used to describe the forces acting on charged particles. It is linked to the potential difference on both ends and to the distance between these measurement points; it is measured in VOLTS/METER (Vm^{-1}), and it is defined as

$$E = \Delta V / L, \tag{1.1}$$

where $\Delta V = V_2 - V_1$ is the potential difference between the two points (1 and 2) between which the electrical field is determined, and L is the distance between the two points.[2]

Building on these definitions it is possible to investigate the relationship between driving forces and resulting flow. If some resistance to the

1. Note that the symbol for the potential (V) is the same as for the unit of measurement.

2. The Greek symbol Δ (capital letter delta) is used to indicate the difference between two values.

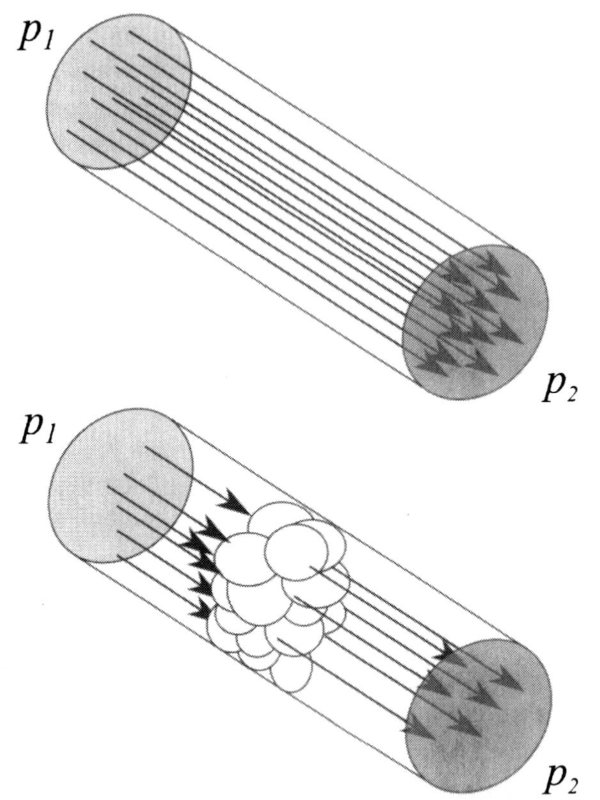

Figure 1.2. Blockage in a hosepipe leads to reduced water flow. This is similar to an electrical resistance.

flow of water molecules is present in the hosepipe—for example, in the form of soil blocking it (figure 1.2)—it is quite obvious that the pressure head between the two ends has to be increased to obtain the same flow. If instead the pressure head remained the same, the flow would decrease. From this consideration one can define the "electrical resistance" as

$$R = \Delta V / L. \tag{1.2}$$

The electrical resistance is measured in OHMS (Ω).[3] When rearranging this definition as (a) $\Delta V = R\,I$ or (b) $I = \Delta V / R$, it becomes apparent that when the resistance R is increased, either ΔV has to be increased propor-

3. The Greek symbol Ω (capital letter omega).

tionally to maintain the same given current I (case a) or, if the potential difference ΔV remains the same, the current I will drop (case b). This is in accordance with the more intuitive considerations made for the water flow above. It also shows the usefulness of expressing such a relationship as a mathematical equation. From one single expression (equation 1.2), different conclusions can be drawn, depending on the most appropriate "viewpoint" for a given question (cases a and b).

When considering the special case where the electrical current is flowing through a piece of metal (e.g., a wire), the electrical resistance R is constant and independent of the current flowing through it. This finding is referred to as Ohm's law and expresses that the current through metal increases proportionally as the potential difference is increased. This is by no means a trivial observation and does not, for example, hold for semiconductors. It is important to distinguish the definition of electrical resistance (equation 1.2), which always applies, from Ohm's law (R = *const.*), which is only true for certain materials and environmental conditions. The two are related, but not identical. It is also worth noting that even in materials for which Ohm's law is applicable, the resistance R may depend on external factors (e.g., on temperature).

1.1.2 Resistivity

Electrical resistance, as introduced above, is a gross measure for a whole sample and depends on the amount of the material that is being measured. A more characteristic property is "electrical resistivity," which is best explained in analogy with bulk density.

When presented with a lump of iron and a ball of cotton, one would commonly consider the former to be "heavy" and the latter to be "light."

Table 1.1. Flow of water as analogy to electricity

	Water Flow	*Electricity*	*Electrical Units*
Carrier	Molecule with mass (m)	Particle with electrical charge (Q)	COULOMB (C)
Transport	Flow rate (n / t)	Electrical current ($I = Q / t$)	AMPERE (A)
Driving Force	Pressure head ($p_2 - p_1$)	Electrical potential difference ($\Delta V = V_2 - V_1$)	VOLT (V)
	Gravitational field (G)	Electrical field (E)	VOLT/METER (Vm^{-1})
Resistance	Flow resistance	Electrical resistance ($R = \Delta V / I$)	OHM (Ω)

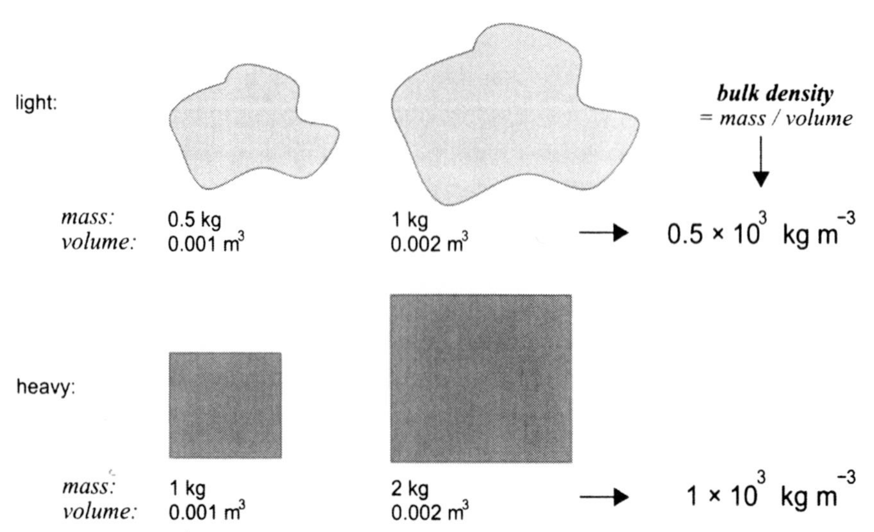

Figure 1.3. Bulk density measures the mass relative to the volume of an object. It can therefore be used to characterize the material and not just the size or amount of a sample.

Even if their masses in grams, as measured with a balance, were the same, this would be our perception. To formalize this impression one introduces "bulk density" as the ratio of mass and volume. Material with a high bulk density has high mass in a small volume (e.g., iron being "heavy"), whereas a low bulk density is characterized by low mass in a large volume (e.g., the cotton ball being "light") (figure 1.3). It is thus possible to characterize and to identify a material according to its inherent property of bulk density rather than through the amount one just happens to have at hand. A reference table of bulk densities can be used to identify the particular material under investigation. This concept of a measure that is specific to the material used, rather than its amount or shape, can also be applied in the electrical case.

In the electrical case the specific electrical resistance is called resistivity and is introduced here using the example of a wire with a constant cross section (figure 1.4). When comparing wires made of the same material but with different lengths, it seems obvious that in a wire of twice the length the electrical current will encounter twice as much resistance. By contrast, if a second wire had the same length but twice the cross section, many more paths were available for the current to travel and the resistance

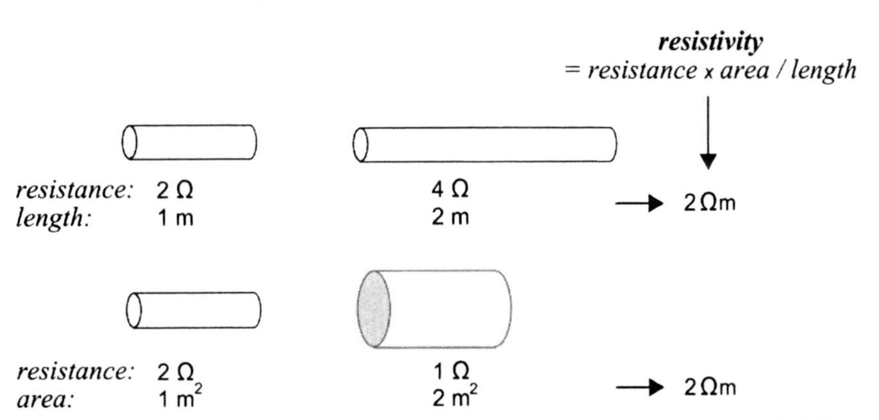

Figure 1.4. For a wire the electrical resistance depends on its resistivity, length, and cross-sectional area. The calculated resistivity is characteristic for the material of the wire.

would therefore be halved. Formalizing these considerations, the following equation can be derived:

$$R = \rho \cdot L / A, \tag{1.3}$$

where ρ is the electrical resistivity, L the length of the wire (in m), and A its cross-sectional area (in m^2). The resistance is hence proportional to the length and inverse proportional to the cross section of the wire and otherwise only depends on the chosen material, which is characterized by its resistivity ρ (the lowercase Greek letter rho). When rearranging equation 1.3 the resistivity of a wire can then be expressed as

$$\rho = R \cdot A / L. \tag{1.4}$$

Its units are Ohm Meter (Ωm).

As stated above, resistivity can be used to characterize materials, and table 1.2 and figure 1.5 provide an overview of the wide range of possible values (Palacky 1987). When looking at the data for clay, soil,[4] and sand, it is clear that the addition of water lowers the overall electrical resistivity

4. Where "soil" is mentioned in subsequent sections, this is meant in the same sense as "ground" and includes sediments as well.

Table 1.2. Electrical resistivity of selected materials

Material	*ρ [Ωm]*
Copper	1.72×10^{-8}
Aluminium	2.83×10^{-8}
Magnetite	6×10^{-3}
Clay (28 percent water)	0.16
Clay (4.4 percent water)	14.5
"Soil" (17.3 percent water)	0.60
"Soil" (3.3 percent water)	16.7
"Sand" (9.5 percent water)	0.95
"Sand" (0.86 percent water)	8.3
Limestone (moist)	4×10^{2}
Limestone (dry)	7×10^{2}
Sandstone	$5 \times 10^{7}...10^{9}$
Granite	$10^{7}...10^{9}$
Glass plate	2×10^{11}
Sulphur (dry)	10^{14}
Fused quartz	$> 5 \times 10^{16}$

of a material, which is due to the low resistivity of water when it carries dissolved salt ions (section 1.3.2).

1.1.3 Conductance and conductivity

While electrical resistance and resistivity describe the impediment to current flow, it is sometimes useful to characterize a material by how well it conducts a current. "Electrical conductance" and "conductivity" are therefore introduced as the inverse of resistance and resistivity, respectively (table 1.3).

Table 1.3. Conductance and conductivity

Quantity	*Definition*	*Units*
Electrical Conductance	$G (= 1 / R)$	Ohm^{-1} or Siemens (Ω^{-1} or S)
Electrical Conductivity	$\sigma (= 1 / \rho)$	Ohm^{-1} Meter^{-1} or Siemens Meter^{-1} (Ω^{-1} m^{-1} or S m^{-1})

1.2 Electrical potentials, fields, and current lines

In the preceding section the concept of electrical resistivity was introduced using the example of a homogeneous metallic wire. Since the material

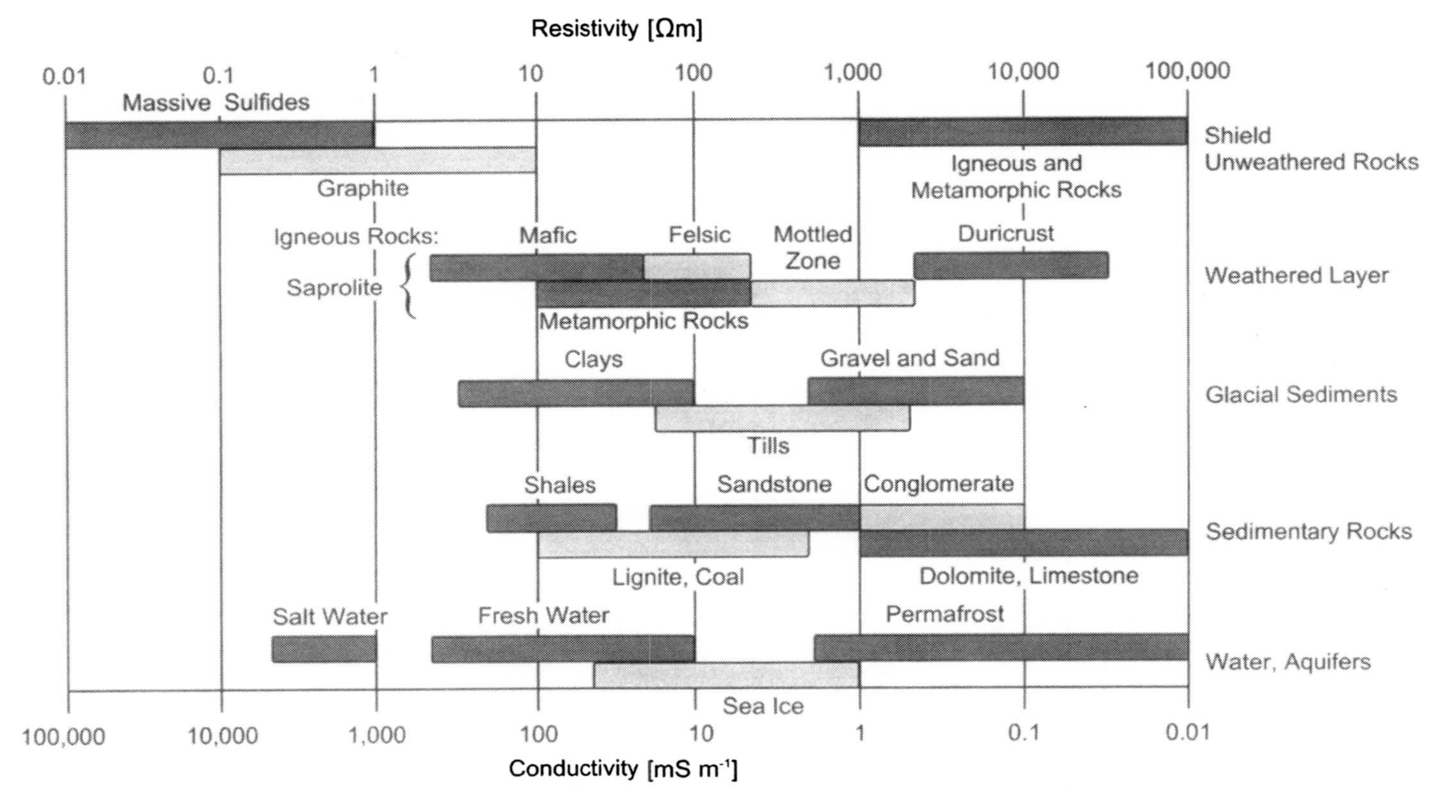

Figure 1.5. Natural materials show a wide range of electrical resistivities (after Palacky [1987]).

property (electrical resistivity) and geometry are the same throughout the object (i.e., having a constant cross section), the relationship between electrical parameters is fairly simple (see equation 1.4). However, in archaeological prospecting electrical current flows through the highly heterogeneous ground, and a more elaborate formalism is required to describe what happens.

It has been mentioned before that the driving force behind electrical current flow is a difference in electrical potentials (similar to the pressure head in the hosepipe with water). To determine the current paths in the ground one therefore needs to know the spatial distribution of these potentials. A voltmeter with one electrode inserted at a fixed location could be used to measure and map the electrical potential at various points in a plane (figure 1.6a). Exactly in the same way as for a topographical contour map, these values can then be joined together to form lines of equal potential value, which are therefore called "equipotential lines" or "equipotentials" (figure 1.6b), so that an appreciation of the distribution of electrical potential can be gained.

If such a diagram were a topographical contour map, water would always run down the steepest slope, perpendicular to the contour lines. The same is true for equipotential lines and electrical currents. The steepness of the topographical slope that determines the flow rate is visualized as the density of contour lines. The equivalent of this topographical gradient is the "electrical field" E that is determined by the change of electrical potential over a certain distance (equation 1.1). If small variations over very short distances are to be considered, it is best to use the differential expression $E = dV / dx$. From these definitions it is clear that the electrical field is high when the gradient of equipotential lines is steep, and it can then drive a high current.

If the electrical potential varies spatially, the resulting current will do so as well. It is therefore insufficient to measure only the total charge that flows through the system ($I = Q / t$), and instead a "current density" j is introduced that quantifies a current that passes through a given cross-sectional area A (figure 1.7):

$$j = I / A. \tag{1.5}$$

The differential expression is $j = dI / dA$, and units are VOLT/METER^2 (Vm^{-2}). To visualize the flow of current in a material one often draws

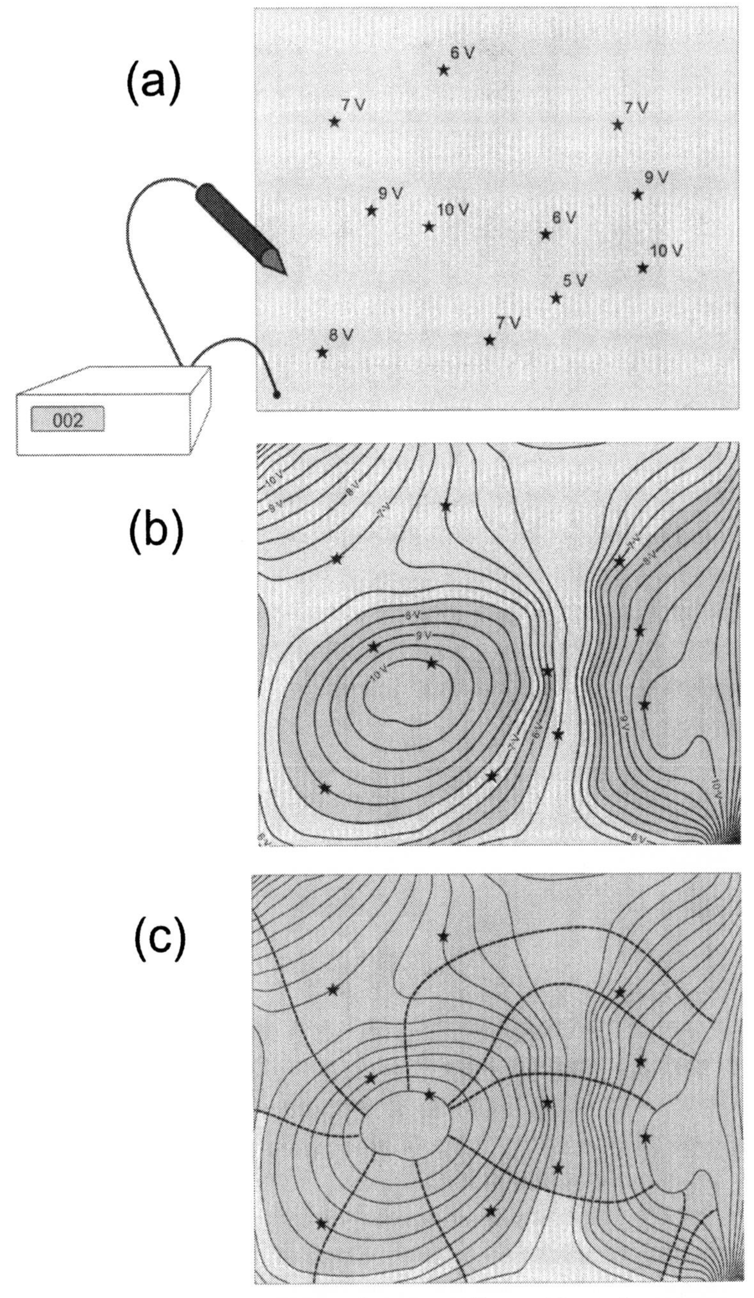

Figure 1.6. Equipotential and current lines: (a) spot measurements of electrical potential in two dimensions; (b) resulting equipotential lines; (c) the current flows perpendicular to the equipotential lines.

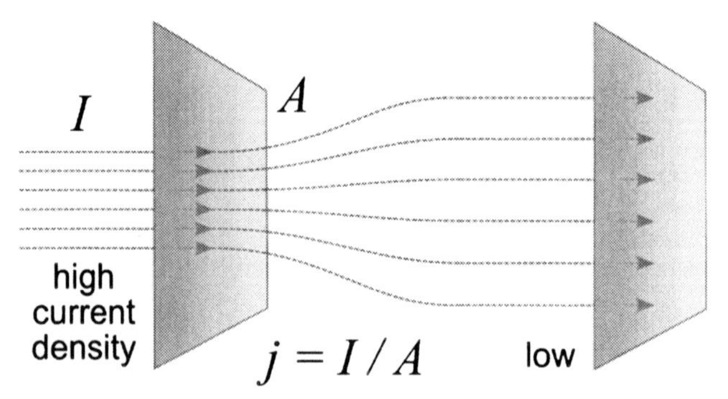

Figure 1.7. Current density is visualized as the density of current lines flowing through a given area.

"current lines" (see figure 1.7). Their total number represents the current I in the system, while their separation represents the current density j. It is important to realize that such a picture is only a symbolic representation; the charged particles also flow between these lines.

To rewrite the resistance definition (equation 1.2) in these new quantities, the special case of a wire (equation 1.3) is combined with the definitions of the electrical field (equation 1.1) and the current density (equation 1.5) to yield

$$E = \rho \cdot j, \tag{1.6}$$

or

$$j = \sigma \cdot E. \tag{1.7}$$

In this expression no reference is made to a specific geometry (unlike for a wire having a constant cross section), and it can be applied in all cases, no matter how spatially complicated they are. This is the reason why these equations, albeit seemingly more abstract, are better suited to describe electrical currents in the ground.

With the introduction of spatially varying electrical quantities it is also possible to consider a resistance to current flow that is caused by a varying geometry. In figure 1.8 a wire with changing cross section causes the current lines to run closer together, symbolizing a higher current density. Since the resistivity, as a material property of the wire, does not

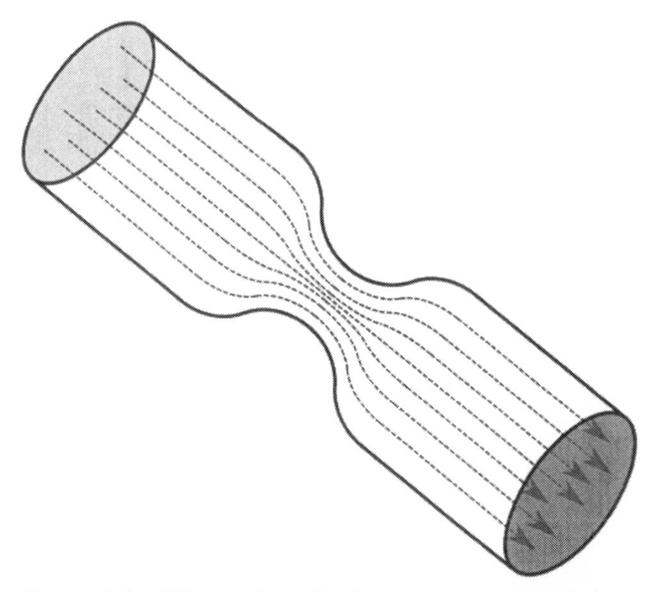

Figure 1.8. Where the wire is narrower current density becomes higher, leading to an increased resistance.

change, it can be concluded (equation 1.6) that the electrical field must have increased in the narrower section. The potential difference measured along this narrow part is therefore also higher, leading to an increased resistance (equation 1.2), since the total current flowing through the material (the number of current lines) remains the same. Independent of the material's resistivity, the varying geometry leads to a change in electrical resistance. This is also evident when looking at the narrow cross section in figure 1.8, confirming that the chosen mathematical descriptions are sensible. In addition, these formalized descriptions allow an explanation of more complex cases, which is the main reason for their use.

It is useful to investigate further the relationship between equipotential and current lines. As the electrical field is the gradient of the equipotential lines (see figure 1.6b), it is always perpendicular to them, like the line of steepest slope is perpendicular to contour lines. Since the electrical field drives current flow and its density (equation 1.6), it follows that equipotential and current lines are perpendicular to each other. This can also be explained by considering that along equipotential lines the potential does not change (that is how they are defined). As a result, there is no force to drive charged particles along them and therefore no current can flow parallel with them. All of the current has to flow away from the equipotential lines, which is "best achieved" if the two sets of lines are perpendicular to each other. It is therefore possible to construct one set of lines if the other is known (see figure 1.6c). For example, currents that flow in the ground very close to the surface cannot flow out of the ground into the air and will therefore flow nearly parallel to the surface. Hence all equipotential lines have to intersect the surface at right angles (figure 1.9). These relationships will be used later to visualize the current flow around features in the ground (section 1.3.2) and to draw conclusions about resulting measurements.

1.3 Earth resistance

The way electrical currents flow through the ground is different from a simple metal wire, and the resulting measurements are usually referred to as "earth resistance," as a reminder of the higher complexity in this case.

1.3.1 Dimensionality

For a wire with constant cross section the measured electrical potential only changes along its length. It is therefore possible to describe the spatial variations of electrical measurements with a single parameter, the position along the wire's length (figure 1.10). This is referred to as a one-dimensional system, even if the wire may be bent into various shapes in a real three-dimensional space. Similarly, if current is confined to a thin sheet of metal or conducting paper, all spatial variations can be described by two parameters (often referred to as "x" and "y," or "easting" and "northing," figure 1.11). In the three-dimensional case of current flowing

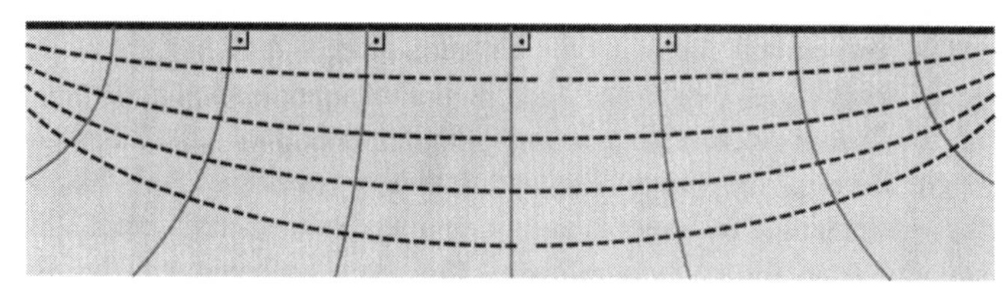

Figure 1.9. Close to a surface the electrical currents have to flow parallel with it. As a consequence, the equipotential lines intersect the surface at right angles.

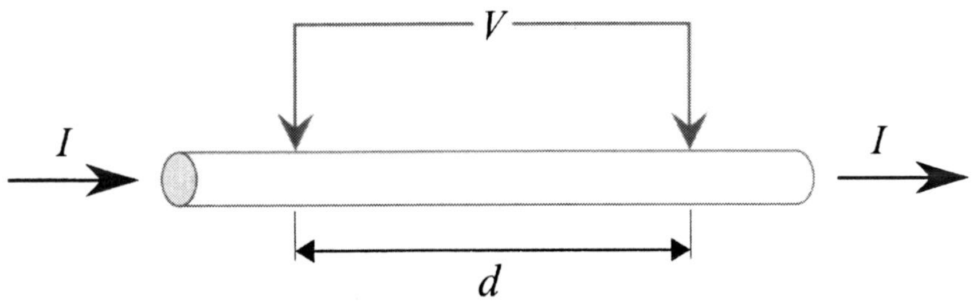

Figure 1.10. The spatial variation of the electrical potential *V* in one dimension (e.g., in a wire) can be characterized by a single spatial parameter—for example, the length *d* along the wire.

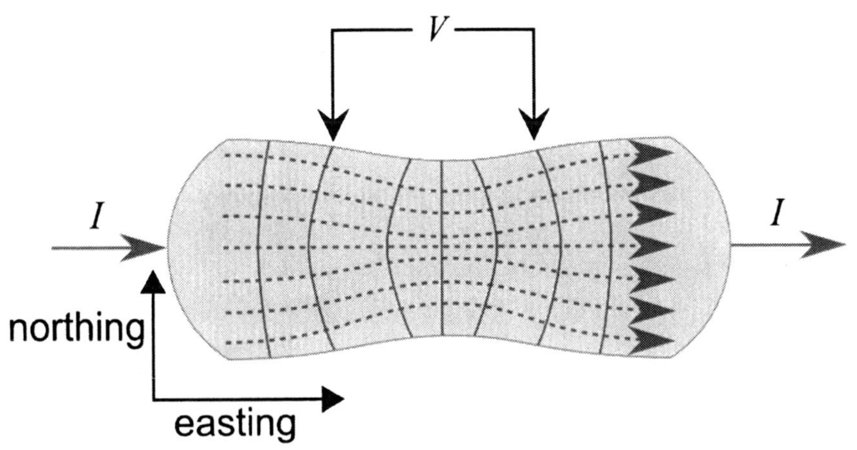

Figure 1.11. In two dimensions spatial variations are described by specifying easting and northing, or x and y.

through a volume of soil, no restrictions on its spatial distribution can be assumed, and earth resistance is inherently three-dimensional (figure 1.12). In this case the equipotential lines become equipotential surfaces in the three-dimensional ground.

1.3.2 Transport of charge

The transport of charge in the ground is fundamentally different from electrical currents in a metallic wire. In metal the free electrons in the conduction band carry the current. They move freely through a metal cable, its connectors, and a battery. There is never a depletion or accumulation of electrons in such a closed circuit. This current flow is termed "electronic."

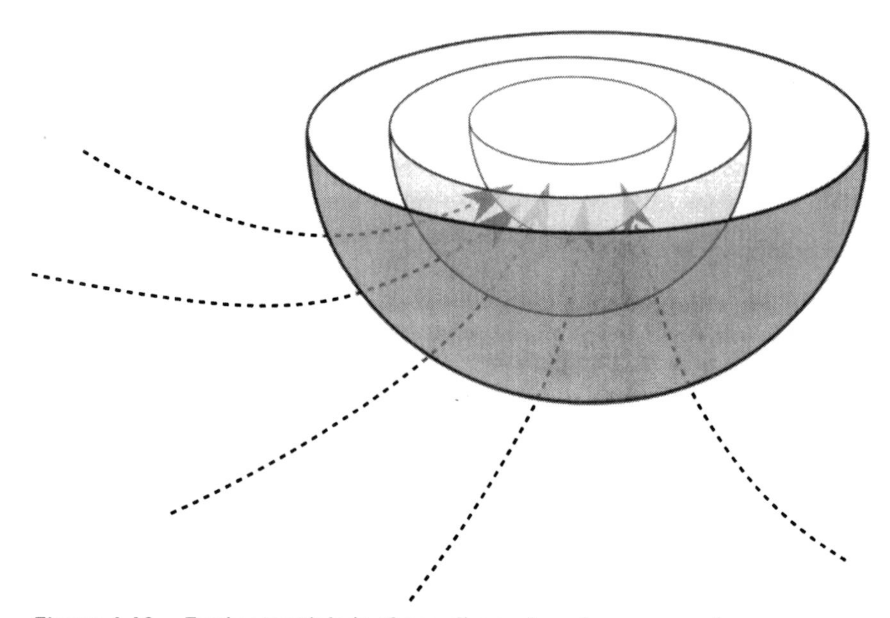

Figure 1.12. Equipotentials in three dimensions become surfaces.

By contrast, no such free electrons are available in the soil. In soil, the current is entirely carried by charged molecules or atoms, called ions. Positive ions (cations) are attracted to a negative electrode in the ground (cathode), while negative ions (anions) are attracted to the positive anode. Where do these ions come from? They are the remains of salt crystals (e.g., NaCl) dissociated in the soil's water (e.g., to Na^+ and Cl^- ions) and therefore depend on the ground's initial composition. It is important to realize that current flow in the earth is related to an electrically driven transport of matter (electrolysis) and is therefore referred to as being "electrolytic."

The electrical resistivity of soils, describing the hindrance to the ions' movements, critically depends on the specific conditions of the ground. First, it is influenced by the availability of salts. While there are some salts in all soils, the degree may vary. Second, and more importantly, it depends on the availability of water. Water is needed to dissolve the salts into their constituent ions and also to facilitate their transport. Soil resistivity is hence mainly governed by the moisture content of the ground. Research into the latter forms the basis of soil science, and a wide range of literature is available (e.g., Cook and Dent 1990; Corwin and Lesch 2003; Michot et al. 2003; Luck et al. 2009; Werban et al. 2009). The major factors influencing soil moisture are pore and grain size, and water content. In addition, resistivity also depends on the mobility of the ions in water, which decreases with temperature and ceases when the water is frozen to ice (Scollar et al. 1990).

Another implication of the electrolytic nature of current flow in the ground is the possible polarization of the electrodes. Since ions cannot escape from the ground they start to accumulate around the respective electrodes, resulting in charged clouds around them. As a consequence of this screening, the force on remaining ions in the ground diminishes rapidly, and the current stops. Even if measures were taken to stop the polarization of electrodes, after a while all ions in the ground would have been transported to the respective electrodes and none would be left to carry further electrical currents. It follows that current flow in the ground can vary with time, and special precautions have to be taken to avoid polarization of the electrodes. This is usually done by reversing the polarity of the current regularly, moving ions alternately in opposite directions and hence avoiding a net transport. In most instruments this alteration is achieved through an electronic switch. The resulting current always has the same magnitude but different polarity, and it is referred to as "switched DC" (DC means "direct current") (figure 1.13). By contrast, an alternating current (AC) would continuously change its magnitude from positive, gradually down to zero, further down to negative, and gradually back again, which causes other effects (e.g., induction) and is not usually used in earth resistance measurements.

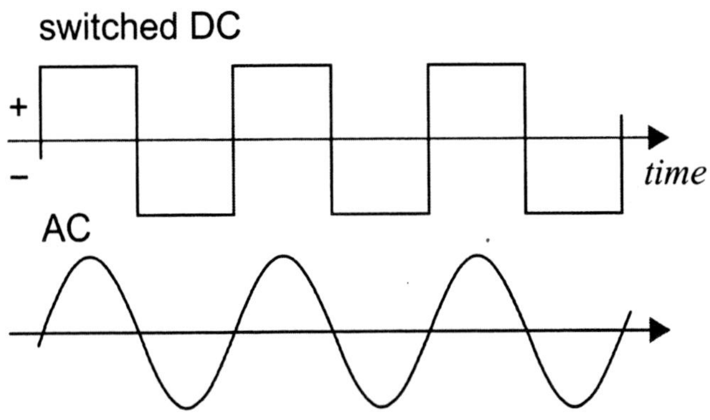

Figure 1.13. A switched DC current only changes its polarity whereas an AC current continuously varies its amplitude.

1.4 Archaeology and earth resistance

In order to use electrical measurements for archaeological prospecting a resistivity contrast must exist between the archaeological features and the surrounding soil matrix. The simplest example is a buried stone foundation. There is usually minimal moisture in such a foundation leading to a high resistivity, whereas the soil around the foundation will normally retain some moisture and have a lower resistivity. This contrast allows detection of the buried foundation with earth resistance measurements from the surface. In the case of the buried stone foundation the contrast is often referred to as being "positive," since the resistivity of the sought-after feature is higher than the background $\Delta\rho = \rho_{feature} - \rho_{background} > 0$. If the foundations were, however, made of bricks in a soil of high ground salinity, the situation might be different. The bricks may be fully saturated with water and salts and therefore have a low resistivity. It will usually still be higher than the surrounding soil (giving rise to a slightly positive contrast), but in extreme cases water-saturated bricks can even have lower resistivity than the surrounding soil due to increased salinity and thereby creating a slightly negative contrast.

Typical features of past human occupation are ditches. After the abandonment of settlements, they become in-filled, for example, through silting or slumping from the sides, and are often no longer visible under current ground cover. However, the fill of a ditch is often more loosely packed, with larger pores for the retention of water, and it normally has a lower resistivity than the surrounding soil. A negative resistivity contrast is therefore typical for buried ditches. However, this is not always the case. As indicated before, the moisture content of soils depends in a complex way on a number of environmental factors, related to water input, retention, and loss (evapotranspiration refers to the loss through sun and wind). With the help of hypothetical weather events some interesting observations can be illustrated.

- In a warm and dry summer, the soil matrix may have dried out considerably and only the ditch retains some moisture. This will lead to a pronounced negative contrast of the ditch.
- If the dry weather continues, the ditch will also lose its moisture and the contrast will gradually disappear, making the ditch undetectable in prolonged dry weather.

- Then it starts to rain heavily, and the large pores of the ditch soak up the water very quickly, even quicker than the surrounding soil with its smaller pore sizes. This leads again to a wetter ditch with a negative contrast.
- The rain continues, and after several weeks all ground is thoroughly wet. By then the contrast will have been lost again and the ditch can no longer be detected with earth resistance measurements.
- Next, sun and strong winds appear and evapotranspiration of the ditch's large pores is much stronger than for the surrounding soil, reducing its water content (i.e., increasing its resistivity) more quickly and leading to a positive contrast.
- As this evapotranspiration continues all soil will become dry and the contrast of the ditch has disappeared again.

These exaggerated and idealized weather conditions help to understand the possible variability of soil moisture and resistivity contrast. While ditches are normally associated with negative resistivity contrasts, situations may occur when they disappear or become positive. Further discussion of water retention in soils can be found in Scollar et al. (1990). An additionally complicating factor is the subsoil geology. Depending on the underlying drainage (e.g., usually good for chalk, poor for clay), further avenues for water loss or retention are open, not just evapotranspiration. It is also worth noting that the example of the ditch (consisting mostly of soil) is particularly complex, and the above example of a stone foundation shows much less variability. However, there have also been cases where a buried stone foundation led to the pooling of water above it, while the surrounding soil was freely draining. For example, in one of my surveys at the landscape park of Dunham Massey the ca. 1.5 m wide foundation of a lost garden statue was rediscovered, showing as a low earth resistance anomaly after a day of heavy rain that had accumulated just above it, under the grass turf.

From this discussion it is clear that some knowledge of recent weather history and soil types is useful for the archaeological interpretation of earth resistance measurements. Several studies have monitored the changing resistivity contrast of ditches over time (Al Chalabi and Rees 1962;

Hesse 1966; Clark 1980; 1996; Cott 1997), and all found that certain times of the year (or weather conditions) are more favorable for the detection of archaeological features than others. In most cases, however, logistical aspects govern survey dates, so little can be done about the prevailing weather. However, at least a record of prior environmental factors should be made to help with the interpretation of results.

Past human habitation has left buried features in the ground that often exhibit a resistivity contrast. The same can be true for geological structures, and sometimes only the spatial layout of the resistivity contrast can help to distinguish between them. The mapping of earth resistance over a sufficiently large area is therefore one of the major techniques used for archaeological prospecting. The following chapters will introduce the concepts in detail and explain what measurements can be expected at the surface if features in the ground exhibit resistivity anomalies.

CHAPTER TWO
RESULTS FROM THE HOMOGENEOUS GROUND

Measured earth resistance depends on many factors, not only on the subsurface resistivity distribution but also on the position of electrodes; in other words, their geometry. To distinguish effects of measurement parameters from anomalies in the data that are caused by buried archaeological features, the simple case of an undisturbed soil will be used to study the geometric effects. This is idealized by the "homogeneous halfspace," a three-dimensional subsurface with constant and isotropic resistivity that extends to infinite depth and is bounded by a horizontal ground surface at the top. Measurements of electrical potential are usually made at this ground level.[1]

2.1 Electrical potential of a single point electrode in the homogeneous halfspace

Before investigating more complex potential distributions in the ground it is important to start with the simple case of a single electrode through which current is injected into the ground. The second current electrode, through which the current has to return to the power supply, is considered to be a very large distance away so as not to interfere with the local potential distribution discussed here. In order to keep the investigation simple, only a "point electrode" is considered, where the contact point

1. When geophysicists look at geological scales they may instead have to consider the case of the earth as a spheroid, which is a much more complicated situation.

between the electrode and the ground is small rather than the electrode being inserted very deeply, like a rod. If the scale is chosen appropriately (i.e., looking at a small metal electrode and several meters of ground), this is a good approximation.

When investigating the symmetry of this arrangement (figure 2.1), it is clear that the current lines have to emerge radially from the electrode since all directions are equal in the homogeneous halfspace. The three-dimensional equipotential surfaces are therefore hemispheres, since they have to be perpendicular to these radial current lines.

To find the relationship between the potential V of one of the equipotential hemispheres, its radius r and the current I through the electrode, one can look at a second, very close equipotential surface with a slightly larger radius $r + dr$, which will have a somewhat bigger potential of $V + dV$ (see figure 2.1). According to the discussion in section 1.1.1, the electrical field between these two hemispheres is $E = dV / dr$. The total current I has to pass through the surface area of the hemisphere $A = 2\pi\, r^2$, yielding a current density of

$$j = I / A = I / (2\pi\, r^2). \tag{2.1}$$

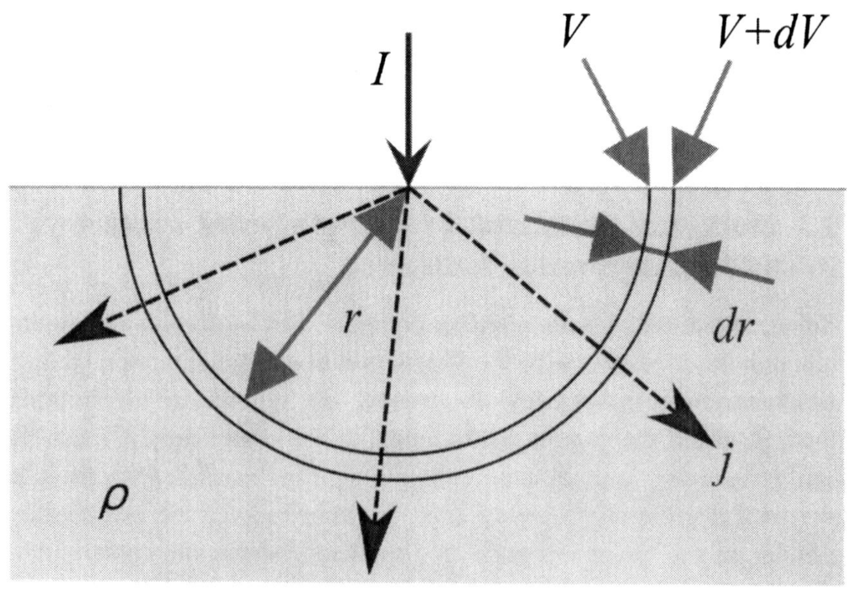

Figure 2.1. The flow of current from a single electrode into the homogeneous halfspace is radially symmetric about that electrode, creating hemispherical equipotential surface.

With the general expression for the resistance definition $E = \rho j$ (equation 1.6) one arrives at

$$dV = \frac{I\rho}{2\pi r^2}\, dr, \tag{2.2}$$

linking the unknown potential difference between the two closely spaced hemispheres to the known parameters. To find the actual electrical potential value V of the hemisphere, the distance to the electrode has to be divided very finely into a number of smaller hemispheres, like onionskins. The individual potential differences dV between them are then noted and all values are added together (figure 2.2). When formalized, this process is referred to as integration.

After such an integration for which the boundary condition is chosen so that the potential at distances extremely far from the current electrode is zero (all current has dispersed there), the resulting expression is

$$V = \frac{I\rho}{2\pi r}. \tag{2.3}$$

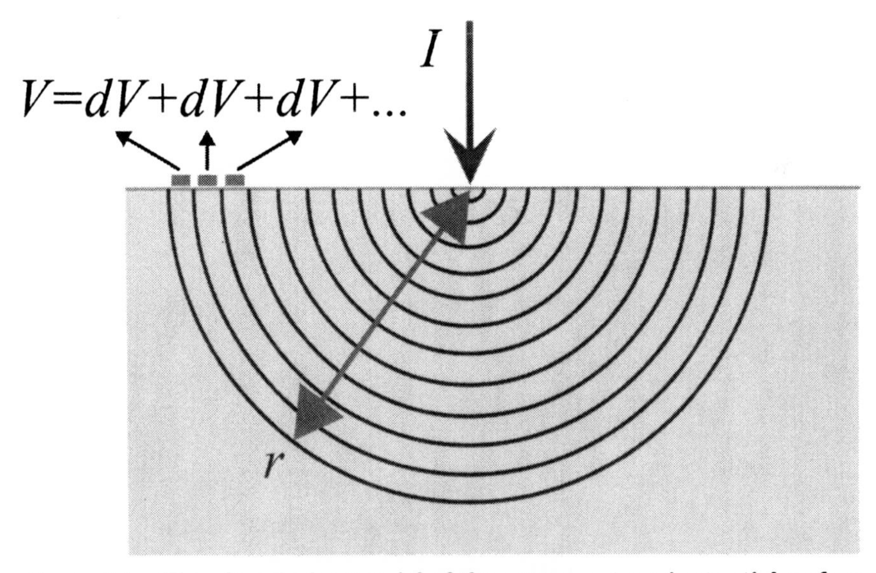

Figure 2.2. The electrical potential of the outermost equipotential surface is calculated by adding (i.e., integrating) the potential differences between all inner surfaces.

The electrical potential at any point in the homogeneous halfspace, including its surface, is therefore inversely proportional to the radial distance from the electrode. The further away a measurement point is from the electrode, the smaller is the potential. This is quite different from the one-dimensional case of the wire, in which the electrical resistance (and hence the potential difference between both ends) *increases* linearly with the length of the wire (equation 1.3).

2.2 Electrical potential of two point electrodes

In many instances the two electrodes through which current enters and leaves the ground are relatively close to each other and require a joint investigation of their combined effect. Fortunately, electrical potentials arising from multiple sources can simply be added together. It is therefore possible to treat the case of two current electrodes as the superposition of an inward (positive) and an outward (negative) current through two individual electrodes separated by a distance L (figure 2.3).

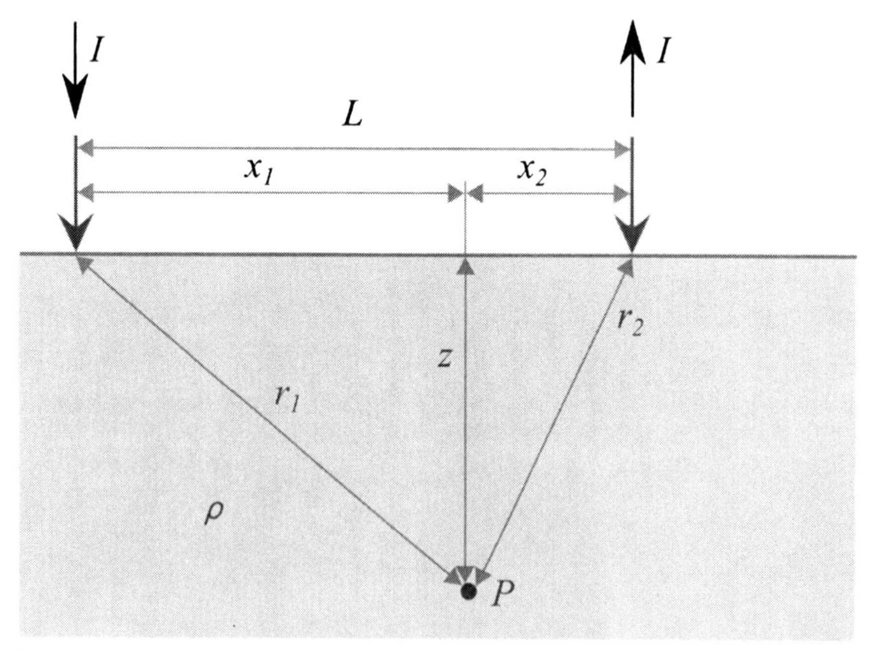

Figure 2.3. Two current electrodes over a homogeneous halfspace are used to calculate the electrical potential at depth z.

For a point *P* in the homogeneous halfspace with the two radial distances r_1 and r_2 from the two electrodes, respectively, the potential is therefore

$$V = \left(\frac{I\rho}{2\pi r_1}\right) + \left(\frac{-I\rho}{2\pi r_2}\right) = \frac{I\rho}{2\pi}\left(\frac{1}{r_1} - \frac{1}{r_2}\right). \tag{2.4}$$

With the point of investigation being at depth z and a distance x_1 and x_2 from the first and second electrode, respectively, the radial distances can simply be expressed as $r_i = \sqrt{x_i^2 + z^2}$, where $i = 1$ or 2 and $x_2 = L - x_1$. The simplest case to investigate is the potential at the ground surface ($z = 0$). There $r_1 = x_1$, $r_2 = x_2 = L - x_1$, and therefore

$$V = \frac{I\rho}{2\pi}\left(\frac{1}{x_1} - \frac{1}{L - x_1}\right). \tag{2.5}$$

For the middle position between the two electrodes $r_1 = r_2$ the potential is $V = 0$. The electrical potential along the connecting line between the two electrodes is plotted in figure 2.4, separately for the two individual

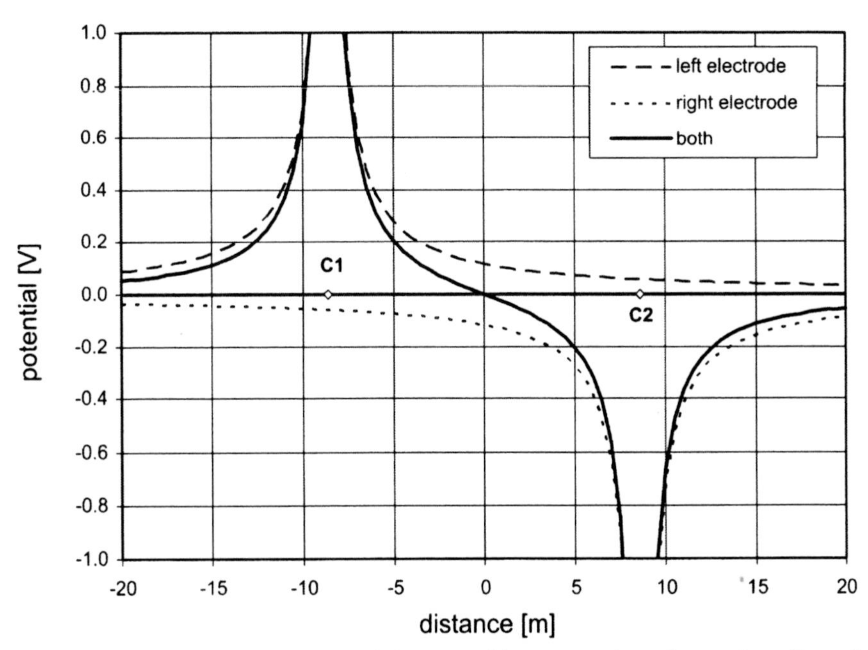

Figure 2.4. The surface potential created by two point electrodes, C_1 and C_2, can be calculated along their connecting line.

electrodes (broken lines) and for the combined electrode arrangement. As would be expected from equation 2.5, the potential rises sharply toward the current electrodes—to a positive level for the inward electrode, to a negative level for the outward electrode, and it is zero in the middle.

While these calculations were made for the homogeneous halfspace, conclusions can also be drawn for measurements over buried features. Such features will distort the current path and will therefore change the electrical potential. From the plot in figure 2.4 it can be seen that an object close to a current electrode would have a strong effect since in that location the potential changes very rapidly and takes on large values. Between the electrodes the results are far less sensitive to such changes in subsurface resistivity. This can also be seen when considering that the electrical field is the gradient of the potential. Based on the gradient shown in figure 2.4, the electrical field is very high at the electrodes (hence showing strong sensitivity to buried features) and lower in the middle.

Equation 2.4 can also be used to evaluate the equipotential lines on the horizontal ground surface (figure 2.5). While they appear spherical in the vicinity of the electrodes, they degenerate to a plane in the center between them. It is also evident that the sensitivity to buried archaeological features decreases considerably as one moves away from the line connecting the two electrodes.

2.3 Current distribution

The distribution and orientation of current flow in the homogeneous half-space can be evaluated similarly, but it requires the use of the mathematical concept of vectors. For a single electrode the current lines emanate radially (see figure 2.1) with a density $j = I/(2\pi r^2)$ (equation 2.1). The case of two electrodes (one for the inward and one for the outward current) can again be constructed from two single electrodes. However, the superposition of current lines is more complex than the simple addition of potential values. Both the strength of current density and the direction of the current flow are relevant. It is therefore necessary to treat the current density as a vector with a length, which is proportional to the strength of the current density, and a direction. Each vector can be expressed through two perpendicular components. The task, therefore, is to find these components for the two current densities, related to electrodes 1 and 2, respectively (figure 2.6).

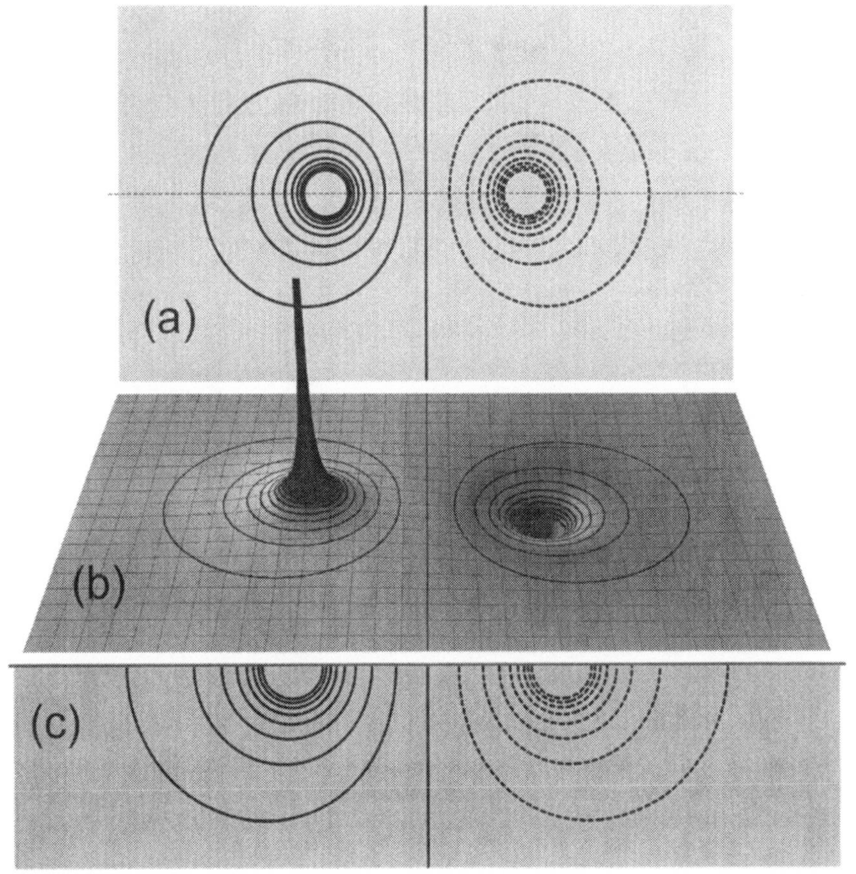

Figure 2.5. The surface potential of two point electrodes can be calculated in two dimensions: (a) horizontal view of equipotential lines; (b) diagram of horizontal equipotential lines with the third dimension showing value of the potential (see figure 2.4); (c) equipotential lines in a vertical section along the connecting line between the two electrodes.

The problem can be resolved geometrically through reference to "similar triangles" (i.e., triangles that look the same; that is, have the same angles, but may be stretched against each other): Triangle $r_1 - z - x_1$ is similar to $j_1 - j_1^{\,z} - j_1^{\,x}$, where j, j^x, and j^z are the strength, horizontal and depth component of the current density vector, respectively.

For these similar triangles the ratios of corresponding sides are

$$\frac{j_i^x}{j_i} = \frac{x_i}{r_i} \; (i = 1 \text{ or } 2), \qquad (2.6)$$

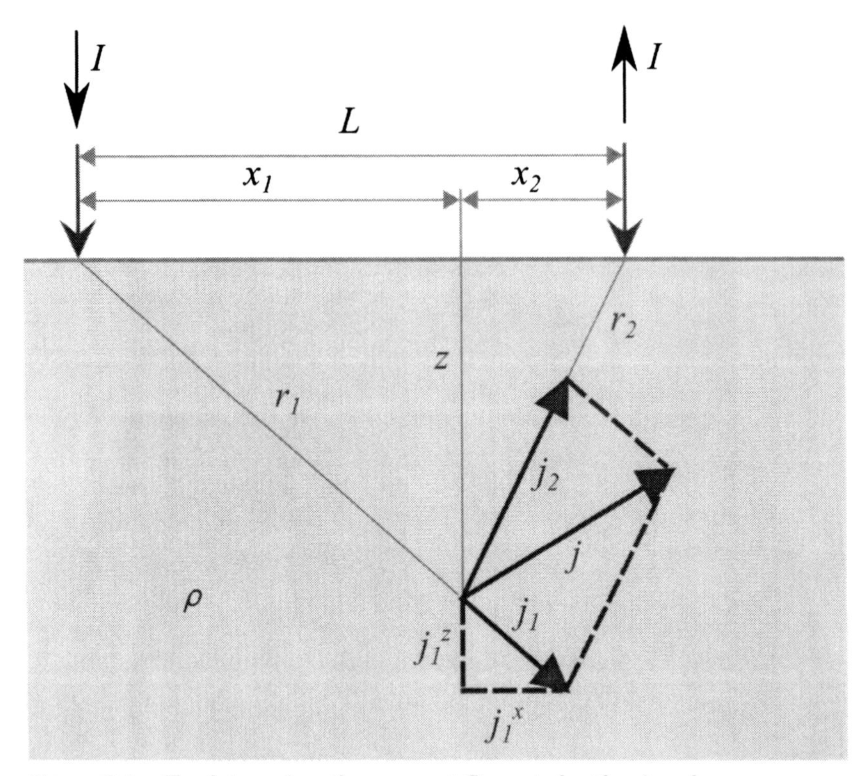

Figure 2.6. To determine the current flow at depth z in a homogeneous halfspace, the two contributing current density vectors j_1 and j_2 have to be added vectorially.

where the radial distance is $r_i = \sqrt{x_i^2 + z^2}$. The x-component of the combined current density (i.e., the superposition of ingoing and outgoing currents) can then be expressed as

$$j^x = j_1^x + j_2^x = j_1 \frac{x_1}{r_1} + j_2 \frac{x_2}{r_2} = \frac{I}{2\pi}\left(\frac{x_1}{r_1^3} + \frac{x_2}{r_2^3}\right). \tag{2.7}$$

This is a fairly complex expression that becomes simpler when it is evaluated for the middle position between the two electrodes where $x_1 = x_2 = L/2$; $r_1 = r_2 = r$. The horizontal component of the current density at depth z between the two electrodes is then given as

$$j^x = \frac{I}{2\pi}\frac{L}{r^3} = \frac{I}{2\pi}\frac{L}{\left(L^2/4 + z^2\right)^{3/2}}. \tag{2.8}$$

When considering the symmetry of the arrangement, it can be seen from figure 2.6 that in the middle between the electrodes the vertical components of the current densities from the two electrodes are identical but opposite, hence canceling each other out. The current flow is therefore entirely horizontal in the middle, and the total current density j is the same as its horizontal component j^x.

There are two ways to interpret equation 2.8. First, one can assume that the two electrodes are fixed and evaluate the change of current density with depth z—for example, using some sort of sensor array buried at various depths. By introducing a convenient dimensionless depth variable $D = z / L$, equation 2.8 can be reformatted into

$$
\begin{aligned}
j &= \frac{I}{2\pi} \frac{L}{\left[\frac{L^2}{4} \left(1 + 4 \frac{z^2}{L^2} \right) \right]^{3/2}} \\
&= \frac{I}{2\pi} L \frac{8}{L^3} \frac{1}{\left(1 + 4D^2\right)^{3/2}} \\
&= j_0 \frac{1}{\left(1 + 4D^2\right)^{3/2}} \quad \text{where } j_0 = \frac{4I}{\pi L^2}.
\end{aligned}
\tag{2.9}
$$

While it comes as no surprise that the current density decreases with depth (figure 2.7, curve (a)), there are a number of points to note. The current density never reaches 0. Instead it decreases continuously, approximately as $1 / (8D^3)$ (since in equation 2.9 the expression $1 + 4 D^2$ can be approximated as $4 D^2$ for very large D). This means that even a deep feature may affect measured results if it has a big enough resistivity contrast to influence a small current density. In addition, the current density at very shallow depth is fairly constant (horizontal start of curve (a) in figure 2.7 for small D). Shallow features with only slightly different depth of burial may therefore produce similar results.

A second way of looking at equation 2.8 is by changing the electrode separation L while keeping the depth of investigation z fixed—for example, by burying a current measuring device at that depth. It is then convenient to introduce another dimensionless quantity $S = L / z$ to characterize the electrode separation. The horizontal current density then becomes

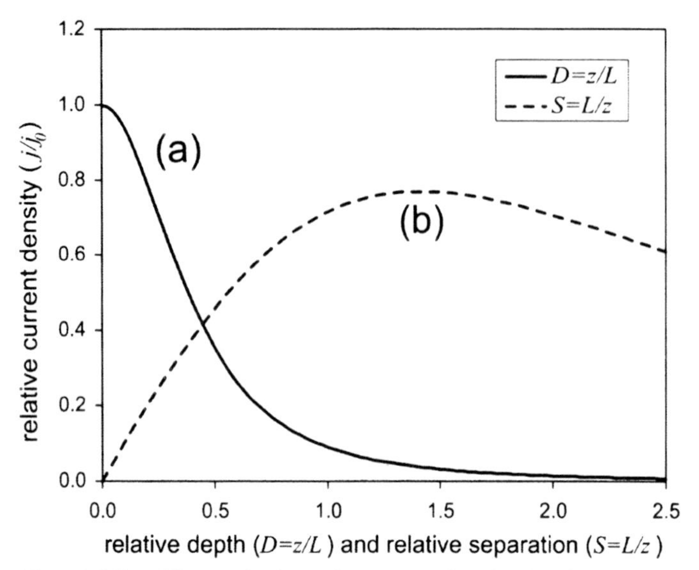

Figure 2.7. The variation of current density in the middle between two electrodes can be calculated with two different considerations: (a) constant electrode separation and varying depth $D = z/L$ **(equation 2.9); (b) fixed depth and varying electrode separation** $S = L/z$ **(equation 2.10).**

$$
\begin{aligned}
j &= \frac{I}{2\pi}\frac{L}{z}z\frac{1}{\left[\left(\frac{L^2}{4z^2}+1\right)z^2\right]^{3/2}} \\
&= \frac{I}{2\pi}S\,z\frac{1}{z^3\left(\frac{S^2}{4}+1\right)^{3/2}} \\
&= j_0^*\frac{S}{\left(\frac{S^2}{4}+1\right)^{3/2}} \quad \text{where } j_0^* = \frac{I}{2\pi z^2}.
\end{aligned}
\tag{2.10}
$$

This is shown in figure 2.7, curve (b), where the ratio of j / j_0^* is plotted (note the difference between j_o and j_o^* in equation 2.9 and 2.10) with a distinctively different shape from curve (a) that has a maximum[2] at $S = \sqrt{2}$. It may seem surprising that the same initial equation (equation 2.8) can yield such different results, depending on the way it is evaluated. This shows again that expressing relationships between variables through mathematical equations can provide new insights. The implication of graph (b) (see figure 2.7) is that when performing investigations with expanding arrays (Vertical Electrical Soundings, see section 3.4.3), a separation exists at which the current density is largest for a given depth.

Another observation can be made from figure 2.6. Between the two electrodes the current densities of the two electrodes point in similar directions, effectively enhancing each other. By contrast, their directions are nearly in opposition on either side of the electrodes, canceling each other out. Thus the current density is mainly confined to the area between the current electrodes.

Combining the considerations for the electrical potential and the current density, it becomes possible to plot a diagram of the vertical equipotential and current distribution in the homogeneous halfspace (figure 2.8, solid and broken lines, respectively). Equipotentials start nearly as hemispheres, not quite centered on the electrode positions. They become larger and larger and degenerate into a vertical plane in the middle between the two electrodes (like a hemisphere with infinite radius). A simplistic way to explain this is to consider that the two sets of hemispheres, which initially form around each electrode (see figure 2.1), push each other apart since they are derived from currents in opposite directions—in and out. The centers of these hemispheres move further and further apart until they eventually degenerate into hemispheres of infinite radius—forming the central vertical plane. The current lines start out radially from the electrodes and become horizontal at the center, as they have to pass perpendicular through this central vertical equipotential plane. Their density decreases with depth and is very low on both sides of the electrodes. This can easily be extended into a three-dimensional picture, and figure 2.5a shows the horizontal plan view of equipotential lines. The current lines are again to be drawn at right angles to them and look similar to those in figure 2.8.

2. The maximum can be found through differentiation of the expression in equation 2.10.

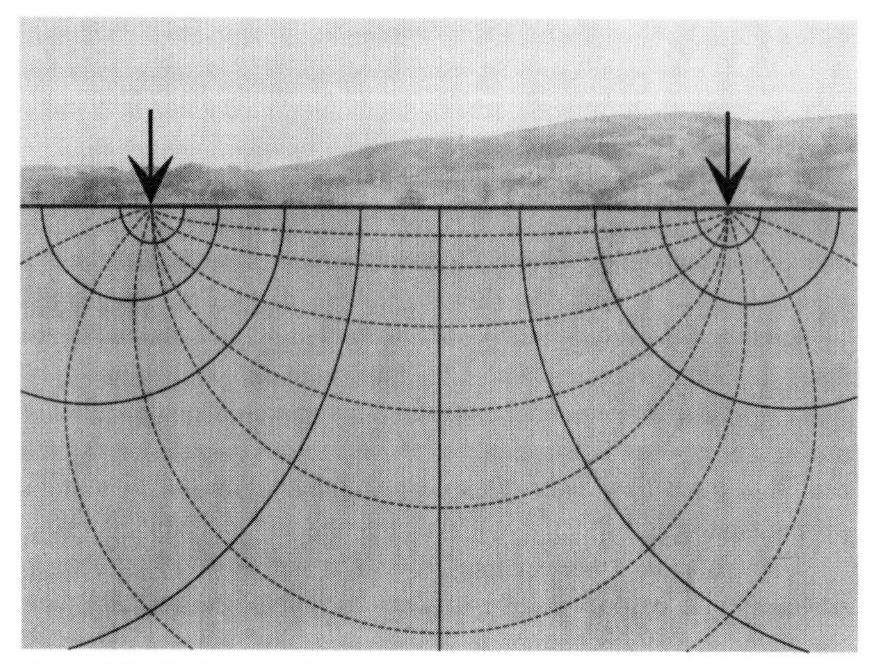

Figure 2.8. Equipotential and current lines in the homogeneous ground (solid and broken lines, respectively).

2.4 Earth resistance arrays

To measure earth resistance four electrodes have to be inserted into the ground: two for the current to enter and leave (labeled C_1 and C_2) and two others to measure the potential difference (P_1 and P_2) (figure 2.9). Other authors have referred to these electrodes as A, B, and M, N, respectively. These four electrodes form the basic constituents of a so-called electrode array that allows the determination of the earth resistance $R = \Delta V / I$, where ΔV is the potential difference measured between P_1 and P_2 and I is the current inserted through C_1 and exiting from C_2. It was shown in previous sections that the distribution of equipotential and current lines, even in the homogeneous ground, is fairly complex, and it is therefore expected that the measured earth resistance crucially depends on the exact position of the four electrodes. The following is a detailed discussion of the results obtained with different electrode arrays over a homogeneous halfspace with resistivity ρ.

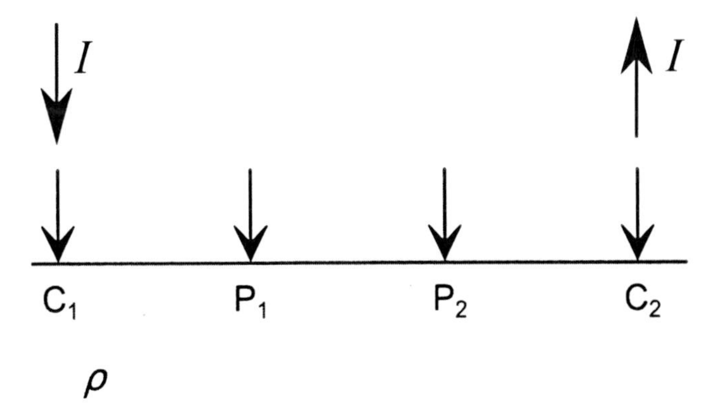

Figure 2.9. A four-electrodes array over a homogeneous halfspace, with electrodes C_1, P_1, P_2, and C_2.

To calculate the measured potential difference, the contributions from the two current electrodes are superimposed. The in-flowing current through C_1 is counted positive; the outward flowing current through C_2 is counted negative. On the surface, the radial distances simply become the distances between the electrodes (indicated below by the notation $[C_1P_1]$, for example). For the first potential electrode P_1, the expression is

$$V_{P1} = \frac{\rho}{2\pi}\left(\frac{I}{[C_1P_1]} + \frac{-I}{[C_2P_1]}\right) \tag{2.11}$$

and the overall potential difference is therefore

$$\Delta V = V_{P1} - V_{P2} = \frac{\rho}{2\pi}\left(\left(\frac{I}{[C_1P_1]} + \frac{-I}{[C_2P_1]}\right) - \left(\frac{I}{[C_1P_2]} + \frac{-I}{[C_2P_2]}\right)\right). \tag{2.12}$$

This is rearranged to yield the corresponding earth resistance

$$R = \frac{\Delta V}{I} = \frac{\rho}{2\pi}\left(\frac{1}{[C_1P_1]} - \frac{1}{[C_2P_1]} - \frac{1}{[C_1P_2]} + \frac{1}{[C_2P_2]}\right). \tag{2.13}$$

To simplify matters, two characteristic array types can be distinguished, depending on the relative location of the first current and potential electrodes.

2.4.1 Type A

In electrode arrays of Type A the potential electrodes lie between the current electrodes. The separation between C_1 and P_1 is taken as the governing geometric unit and labeled α (figure 2.10). All other distances are expressed relative to it ($P\alpha$ and $B\alpha$). While α is a distance measured in meters, the relative factors P and B are purely numeric and have no units.

With these conventions it is possible to rearrange equation 2.13 to yield a simplified expression for the resistance of

$$R = \frac{\rho}{2\pi\alpha}\left(1 - \frac{1}{P+B-1} - \frac{1}{P} + \frac{1}{B}\right). \tag{2.14}$$

In this equation the absolute values of resistivity and the geometric unit α have been grouped together, and the bracket only contains information about the relative location of the electrodes, only using unitless relative factors. This accounts for the fact that a scaling of the whole array (changing α) will not alter its spatial characteristics (determined by the factors in the bracket).

Wenner array

The first example of such a Type A array is the so-called Wenner array, where all four electrodes are equidistant and in a line with a separation between them of a (in meters, figure 2.11).

Comparing figure 2.11 with the general diagram in figure 2.10, it is clear that $\alpha = a$, $P\alpha = 2a$ (hence $P = 2$), and $B\alpha = a$ (hence $B = 1$). Inserted into equation 2.14 this yields

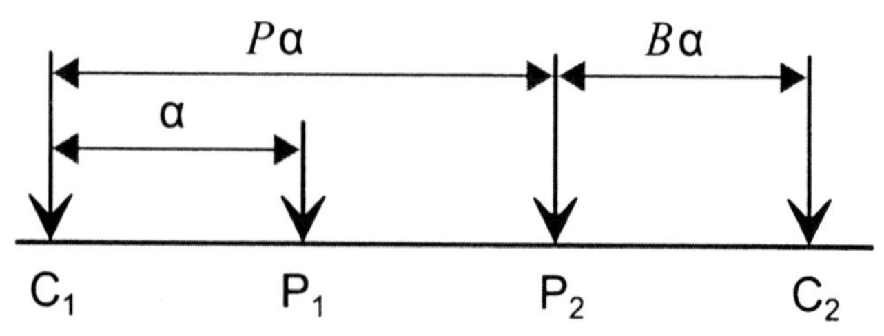

Figure 2.10. Type A collinear electrode array.

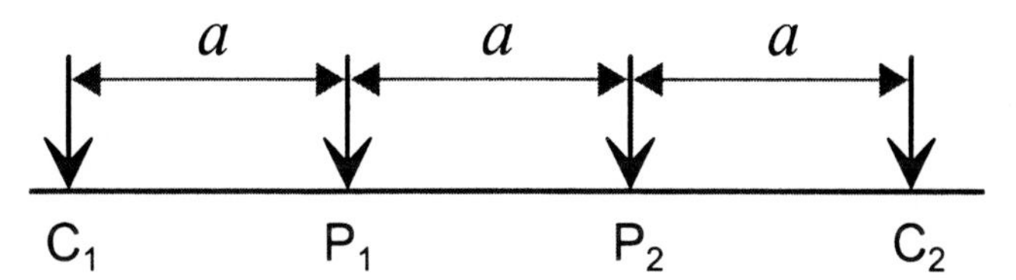

Figure 2.11. Wenner array.

$$R = \frac{\rho}{2\pi a} \tag{2.15}$$

for the earth resistance measured with the Wenner array.

Twin-probe array

In a twin-probe array the order of the electrodes is the same, but the first pair (C_1P_1 with separation a, the "mobile electrodes") is a large distance away from the second pair (C_2P_2 with separation b, the "remote electrodes," figure 2.12). The two sets are connected via a long cable. In the ideal case the distance can be considered as infinite (symbol ∞).

In a twin-probe array the parameters become $\alpha = a$, $P = 1+\infty \approx \infty$ (since 1 can be neglected compared to the large distance) and $B = b / a$. As a result, the earth resistance of a twin-probe array is

$$R = \frac{\rho}{2\pi a}\left(1 + \frac{1}{B}\right), \tag{2.16}$$

since $1 / \infty$ is approximately zero. In many cases the electrode separation of the two pairs is similar ($B = b / a \approx 1$), which leads to the often-used simplified expression

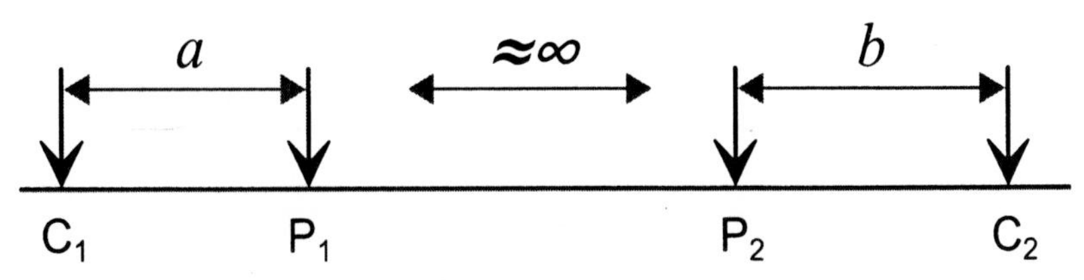

Figure 2.12. Ideal twin-probe array.

$$R \approx \frac{\rho}{\pi a}. \tag{2.17}$$

Comparing this result with the Wenner array, it is apparent that just by moving the electrode pairs far apart from each other, while maintaining their separation *a*, the measured earth resistance doubles. This highlights the importance of these considerations. Even over the homogeneous half-space the measured earth resistance can vary considerably if the arrangement of electrodes is changed.

For a twin-probe array where the pairs of electrodes are at a considerable distance, but not infinitely apart (figure 2.13), the calculation has to take the finite difference Ca between P_1 and P_2 into account. For a simpler calculation, only the case will be considered here where the remote electrodes have the same separation as the mobile electrodes ($B = 1$).

In the case of a twin-probe array with finite distance between mobile and remote electrodes, $\alpha = a$, $P\alpha = a + Ca$ (hence $P = C + 1$) and $B\alpha = a$ (hence $B = 1$). This results in

$$R = \frac{\rho}{\pi a} \frac{C}{(C+1)}. \tag{2.18}$$

If C is considerably bigger than 1, this expression is approximated by equation 2.17, confirming that the latter represents the limit for very large distances between the electrode pairs. Comparing the electrode arrangement with figure 2.11, it can be seen that the finite twin-probe array becomes the same as a Wenner array for a close distance between the electrode pairs with $Ca = a$ (i.e., $C = 1$). Equation 2.18 then converges to the Wenner expression equation 2.15.

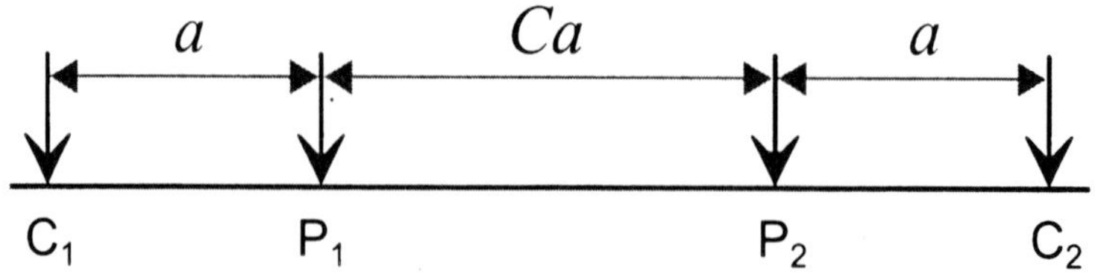

Figure 2.13. Twin-probe array with finite distance between mobile (C_1, P_1) and remote electrodes (P_2, C_2).

Schlumberger array

Another electrode arrangement that is sometimes used is the Schlumberger array. There the two current electrodes are a far distance b apart, and the potential electrodes are closely spaced (separation a) in the center of the array. The whole arrangement is symmetric relative to its center (figure 2.14).

The array parameters for the Schlumberger array can again be deduced from the diagram yielding $\alpha = (b - a) / 2$, $P = (b + a) / (b - a)$, and $B = 1$, hence

$$R = \frac{\rho 4a}{\pi(b^2 - a^2)}. \tag{2.19}$$

If the separation of current electrodes is much bigger than the potential electrode separation ($b \gg a$), this can be simplified to

$$R \approx \frac{\rho 4a}{\pi b^2}. \tag{2.20}$$

2.4.2 Type B

For the second type of linear arrays both current electrodes are on one side of the array, and the potential electrodes are on the other side. As before, the separation between C_1 and P_1 is referred to as α (figure 2.15) but now spans the whole length of the array.

For this electrode arrangement the expression for the measured resistance becomes

$$R = \frac{\rho}{2\pi\alpha}\left(1 - \frac{1}{1 - P + B} - \frac{1}{P} + \frac{1}{B}\right), \tag{2.21}$$

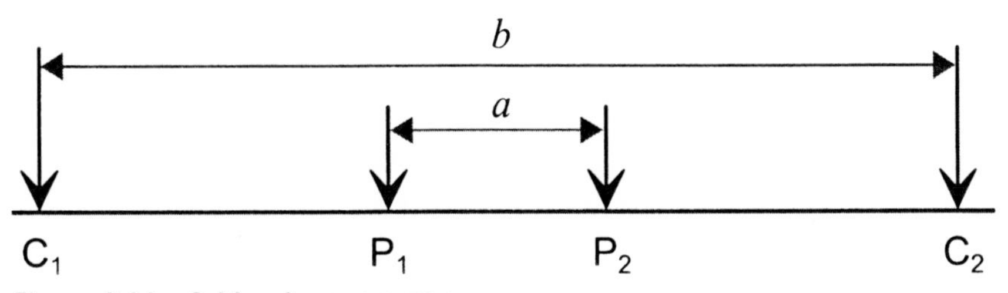

Figure 2.14. Schlumberger array.

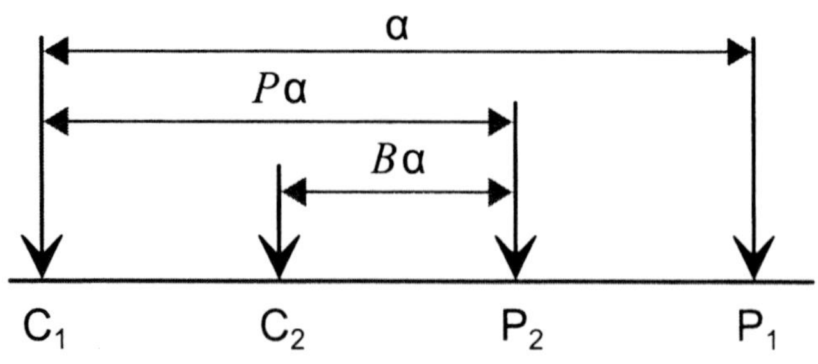

Figure 2.15. Type B collinear electrode array.

where only the signs under the first fraction have changed compared to equation 2.14.

Double-dipole array

If the four electrodes of a Type B array are equidistant, the array is referred to as an equidistant double-dipole array (sometimes called Wenner-β, figure 2.16; the Wenner-γ array has an electrode arrangement of C_1-P_1-C_2-P_2).

In such a double-dipole array $\alpha = 3a$, $P = 2/3$, and $B = 1/3$, yielding

$$R = \frac{\rho}{6\pi a}. \tag{2.22}$$

On first sight it may be surprising that with the same electrode separation as for the Wenner array the resistance drops to one-third (compared to equation 2.15) if the order of the electrodes is changed. However, with the Wenner array the potential difference is measured between the current electrodes where current density was shown to be high, and so the electrical field and potential differences are high. By contrast, the double-

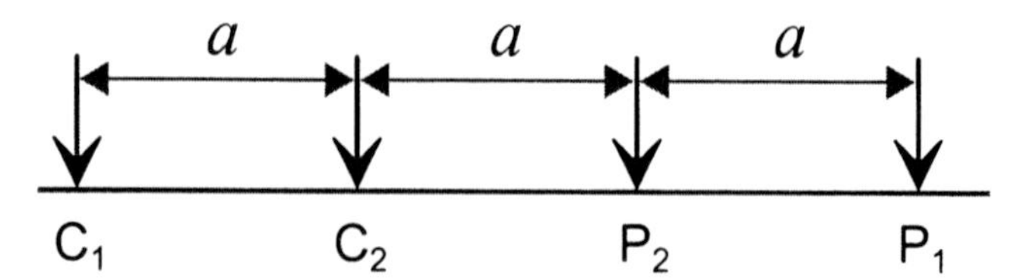

Figure 2.16. Equidistant double-dipole array.

dipole array measures potential differences outside of the current electrodes where the current density is low, yielding a lower earth resistance.

In a general double-dipole array, the distance between the C_1C_2 and P_1P_2 electrodes may vary (figure 2.17).

In such a general double-dipole array, the array parameters become $\alpha = a\,(C + 2)$, $P = (C + 1) / (C + 2)$, and $B = C / (C + 2)$, and after a lengthy arithmetic calculation the earth resistance is found as

$$R = \frac{\rho}{\pi a} \frac{1}{C(C+1)(C+2)}. \tag{2.23}$$

For $C = 1$ the array resembles the equidistant case and, as expected, equation 2.23 becomes the same as equation 2.22 above.

2.4.3 Square arrays

The four electrodes of an array do not have to be arranged along a line. Another very useful configuration is the square array, whereby the electrodes form the corners of a square with a side length of *a*. This is a very compact arrangement that allows the measurement of earth resistance for the volume of soil underneath the square. There are two basic arrangements of electrodes: the pairs of current and potential electrodes (e.g., C_1C_2) can be arranged either along the sides of the square (separation *a*, figure 2.18a) or across the diagonal (separation $\sqrt{2} \cdot a$, figure 2.18b). These two configurations are usually referred to as α and γ, in accordance with the way an "angled" Wenner array would look (namely, C_1-P_1-P_2-C_2 and C_1-P_1-C_2-P_2, respectively). Rotating the α configuration by 90° results in the β configuration C_2-C_1-P_1-P_2. The difference between the α and β configurations is merely in relation to the direction of a traverse over a feature. For the α configuration the electrode pair C_1-P_1 is parallel to the

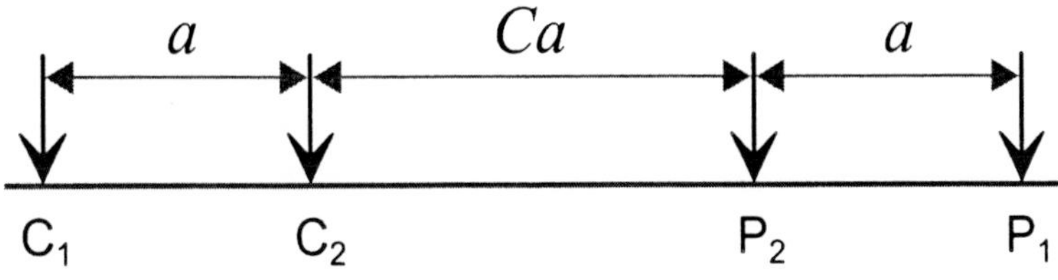

Figure 2.17. General double-dipole array.

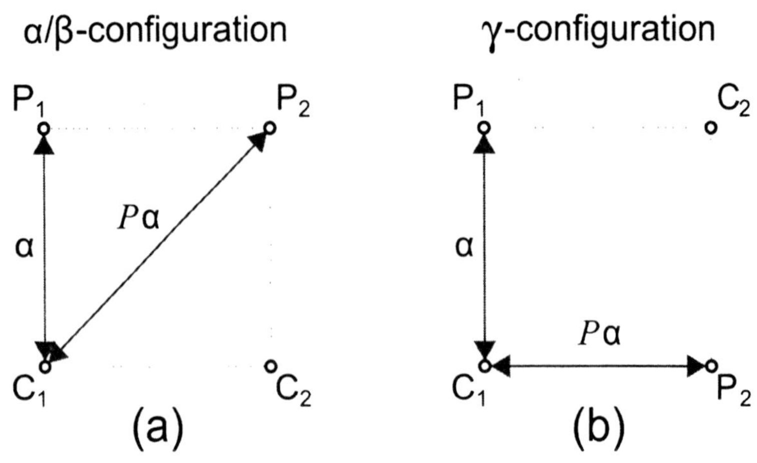

Figure 2.18. Square-array: (a) α/β-configuration; (b) γ-configuration.

direction of the traverse; for the β configuration the electrode pair C_1-C_2 is parallel to that direction[3] (Aspinall and Saunders 2005).

For the α configuration $\alpha = a$, $P = \sqrt{2}$, $B = 1$, and one calculates

$$R = \left(2 - \sqrt{2}\right)\frac{\rho}{2\pi \cdot a}, \tag{2.24}$$

using equation 2.14 for the Type A arrangement. Interestingly, the γ configuration yields $\alpha = a$, $P = B = 1$, and therefore $R_\gamma = 0$. This can also be seen from a direct inspection of the electrodes, as P_1-P_2 forms a line that is perpendicular to, and in the middle of, the line connecting the current electrodes C_1-C_2. It was shown before that the electrical potential everywhere on that line is 0, leading to $R_\gamma = 0$. However, if the ground is not homogeneous, this will lead to an easily detectable deviation of the potential difference from 0 V that is either positive or negative. Since data are usually recorded as earth resistance ($R = \Delta V / I$) this can, perhaps surprisingly, result in negative earth resistance values. Using some basic mathematical manipulations, it can be shown that the earth resistance measured with the three different square array configurations are related as

---

3. Note that this definition of α and β configuration is not consistent in the literature.

Figure 2.19. Wheels of the MSP40 sensor platform act as a square array.

$$R_\gamma = R_\alpha - R_\beta \qquad (2.25)$$

(Aspinall and Saunders 2005). For the homogeneous ground the rotation of the array has no influence and α and β measurements are identical, confirming that $R_\gamma = 0$. For a real earth resistance survey, however, these three measurements can be used to assess the homogeneity of the ground (Tsokas et al. 1997), and modern systems therefore use multiplexed switches to record at least two of them for each measurement location.

The square array was initially introduced to archaeology as a simple table with four spiked legs by Tony Clark (1968) (see also figures 37 and 38 in Clark 1996) and was soon used for wheeled devices, whereby the four spiked wheels of a cart form the four electrodes of the array (Hesse et al. 1986). Today the Geoscan MSP40 is a hand-drawn example of such a wheeled device (figure 2.19), particularly suitable for sensitive soil cover, while Geocarta's ARP is towed by a quadbike at high speed. The latter device has several pairs of wheels to record earth resistance for different soil depths (Dabas et al. 2012).

CHAPTER THREE
THE ELECTRICAL SIGNATURE OF FEATURES IN THE GROUND

So far, only the simple case of homogeneous ground has been discussed, and it was found that the recorded earth resistance depends strongly on the electrode array used for its measurement. In this chapter, the more realistic case of resistivity variations in the ground will be investigated. These may be due to either buried archaeological features or geological structures. Two cases will be considered in detail: the "buried sphere" represents a localized feature (e.g., a pit), and the "layered earth" helps to understand results obtained over extended features (e.g., platforms, floors, or topsoil/subsoil boundaries). More generalized features will be discussed at the end of the chapter.

3.1 Apparent resistivity

When considering the homogeneous halfspace, relationships between the measured earth resistance and the ground resistivity were found. These depended only on the arrangement of the electrodes that were used for the injection of the current and the measurement of a potential difference. In reality, the ground is heterogeneous and has different resistivities in different locations and at different depths; the situation is hence considerably more complex than the homogeneous case described before. From figure 3.1 it is clear that a single measurement of earth resistance made at the surface must be a complex function of all the different resistivities in the ground underneath. The surface measurement only provides a summarized

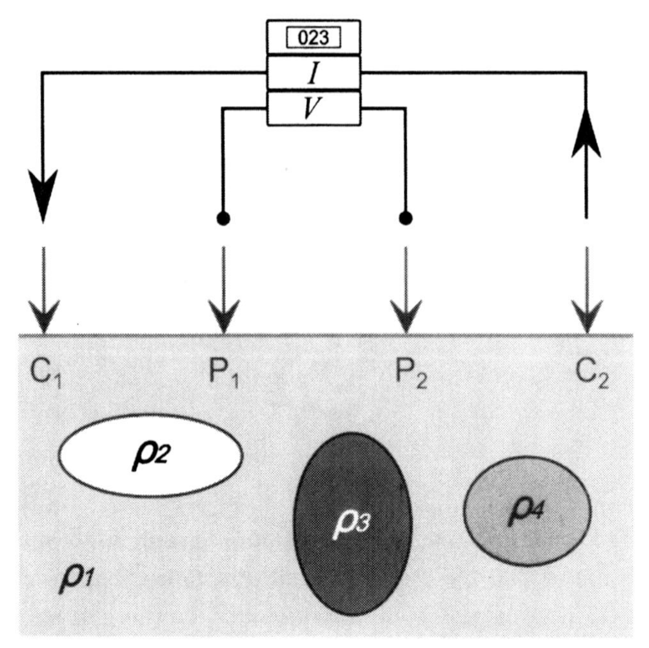

Figure 3.1. A single earth resistance measurement is influenced by the heterogeneous resistivity distribution in the ground.

view of the ground. It would be desirable to deduce the distribution of all subsurface resistivities from earth resistance measurements on the ground surface. However, there is no unique solution to such an "inversion" of earth resistance measurements since several different resistivity assemblages would produce nearly identical earth resistance measurements at the surface. This difficulty is inherent to the investigated problem—namely, trying to deduce three-dimensional soil values from only two-dimensional measurements on the surface. Sufficient information is simply not available. The same is true for most other geophysical measurements—for example, magnetometer surveys (Aspinall et al. 2008).

It was shown in section 2.4 that even over a homogeneous halfspace, different electrode arrays lead to different values of recorded earth resistance. To compare results from various arrays and to separate the effect of the heterogeneous ground from variations introduced by the geometrical arrangement of electrodes, the concept of "apparent resistivity" is introduced. Simply speaking, the apparent resistivity is the resistivity that a homogeneous ground would need to have so as to produce the recorded earth resistance value. In other words, it is "some sort of average" of all the soil resistivities underneath an electrode array.[1] Given an earth resistance measurement R made with a certain electrode array, it is possible to calculate an apparent resistivity ρ_A in such a way that for the simple case of a homogeneous halfspace the apparent resistivity will be identical to the true resistivity ρ of the halfspace.

Generally, apparent resistivity is taken as $\rho_A = K\,R$, where K is a geometry factor related to the arrangement of the electrode array and has the dimension of a length and R is the measured earth resistance. To reproduce results obtained in section 1.1 for the one-dimensional wire one has to choose $K = A / L$ (see equation 1.4). In the discussion of electrode arrays in section 2.4 a common factor of $2\pi\, a$ occurred in most equations, and the geometry factor is hence often written as $K = n\, 2\pi\, a$. The apparent resistivity related to an earth resistance measurement can consequently be expressed as

$$\rho_A = n \cdot 2\pi a R, \tag{3.1}$$

where a is the characteristic electrode separation of the chosen array and the array index n is a numerical factor without dimensions that depends on the arrangement of electrodes. To find the value of n for different arrays, equation 3.1 is compared with equations 2.15 to 2.23, and some results are listed in table 3.1.

It has to be stressed that apparent resistivity is merely a different way of expressing the measured earth resistance. Only for the homogeneous ground is it identical to the ground resistivity. In all other cases it is similar to an "average" of the varied subsurface resistivities and cannot be attributed to one single point in the ground. Nevertheless, it is a useful quantity to relate measurements on the surface to electrical properties underground. For example, measurements made with different electrode separations can be compared more easily, accounting for the purely geometric variations (see section 2.4). Similarly, results from different electrode arrays can more easily be compared. However, it will be shown in subsequent discussions that a certain dependency on array geometry will remain even if all measurements are expressed as apparent resistivities.

---

1. "Average" is used here in a figurative and not mathematical sense.

Table 3.1. Array index *n* for common electrode arrangements

Array	*Array index n*
Wenner	1
ideal twin-probe	0.5
equidistant double-dipole	3
general double-dipole	$0.5\ C\ (C + 1)\ (C + 2)$

Note: For the general double-dipole, the relative dipole distance C is given in figure 2.17 and by equation 2.23.

It is important to remember that in the first instance all electrical measurements record differences in electrical surface potential and are hence expressed as "earth resistance." They cannot usually be attributed to a single resistivity value of the heterogeneous ground and it is therefore impossible to make direct resistivity measurements, unless, of course, the volume of soil is so small that it can be considered homogeneous for the purpose of an archaeological investigation (e.g., a cube of 0.05 m). Earth resistance measurements should therefore not be referred to as "resistivity measurements."

3.2 Current distribution around buried features

A series of earth resistance measurements along a line on the surface is often referred to as a transect. If it passes over an area with subsurface resistivity contrast, it is expected that the measurements change from their background value as they approach the buried feature and revert back to the background levels on the other side. Before these changes will be investigated in detail, some considerations about the possible current flow in the ground are necessary. In figure 2.8 the current and equipotential lines were drawn for the homogeneous ground. Next, it will be explored how they change if a localized conducting or insulating object is introduced, respectively.

The potential difference between any two points on the surface of a *perfect* conductor ($\rho = 0\ \Omega\text{m}$) is 0 since ΔV is proportional to R, which is 0. The conductor forms an equipotential surface, and current lines must intersect it at right angles. Equipotential surfaces in the vicinity of this feature will be of similar shape, a bit like onionskins, due to their proximity.

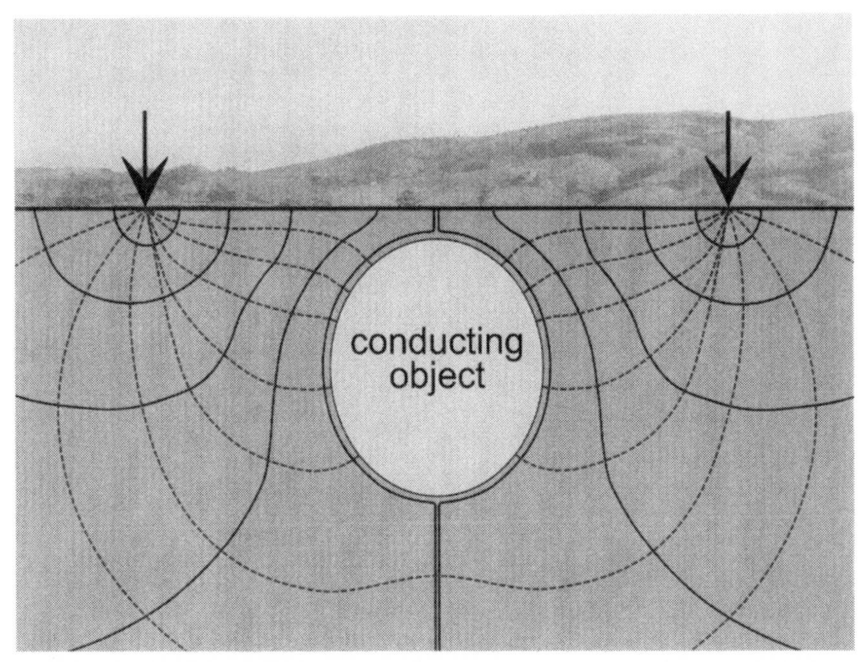

Figure 3.2. Equipotential and current lines around a perfectly conducting object (solid and broken lines, respectively).

The overall picture (figure 3.2) shows how current lines are attracted to the conducting feature, while equipotential lines are pushed away from it.

By contrast, for a perfect insulator ($\rho = \infty$) no current can flow into the object; all of it has to flow around the object. Therefore the current lines have to bend around the outer shape of the feature, while the perpendicular equipotential lines intersect the surface at right angles. Overall (figure 3.3), the current lines are pushed aside, and the equipotential surfaces are attracted to the feature.

Considering that earth resistance measurements are normally taken from the ground surface, some important observations can be made. While the changes to current lines and equipotential lines may be quite pronounced at depth (see figures 3.2 and 3.3), only small changes will "reach" the surface. Only these small changes are measured, and properties of buried features then have to be estimated from those.

For the conductor the equipotentials are pushed apart and at the surface become less dense over the feature, leading to a lower potential

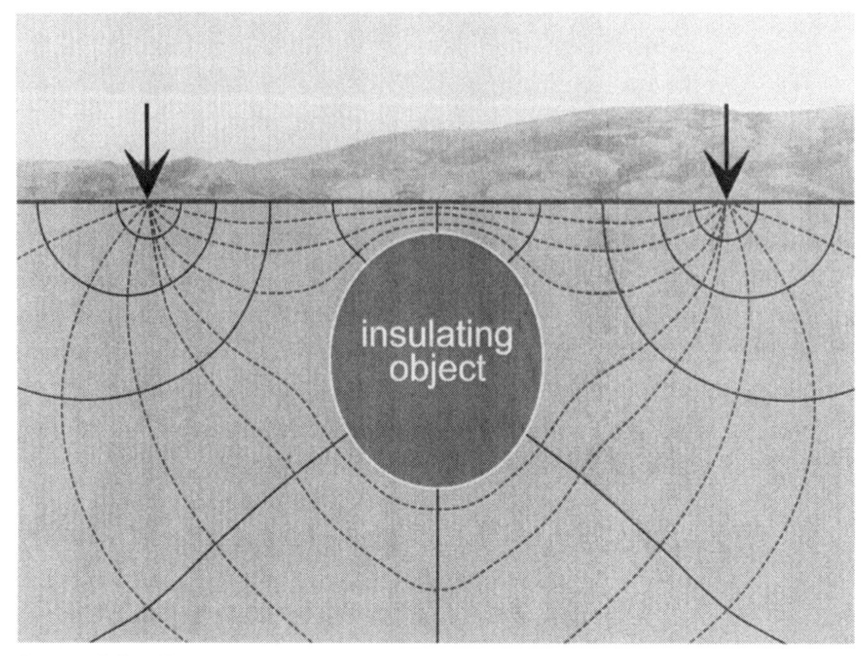

Figure 3.3. Equipotential and current lines around a perfectly insulating object (solid and broken lines, respectively).

difference at a given electrode separation. Hence the measured earth resistance is reduced. The opposite is true for the perfect insulator, where the equipotential lines tend toward the feature and become denser at the surface. The potential difference and earth resistance increase as a result.

3.3 Current flow at boundaries

The previous discussion referred to perfect conductors (ρ = 0 Ωm) and insulators ($\rho = \infty$) and exploited the orthogonality of current line and equipotential lines. To investigate the change of these lines at boundaries between areas with two different resistivities (ρ_1 and ρ_2), a closer examination is necessary.

For a horizontal boundary along the x direction (figure 3.4), the following boundary conditions have to be met:[2]

2. This follows from Maxwell's equations of electromagnetism.

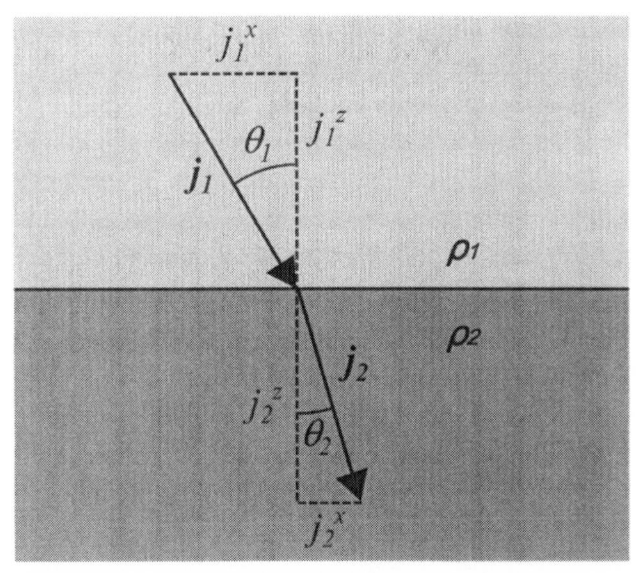

Figure 3.4. Current lines are refracted at a horizontal boundary, shown in vertical view (x/z plane).

$$j^z = const. \text{ and } E^x = const. \tag{3.2}$$

This means that the vertical (z-) component of the current density has to be the same on both sides of the boundary. In other words, current that flows straight in on one side has to flow out on the other, and what flows along the boundary is irrelevant. For the electrical field, by contrast, the horizontal (x-) component has to be identical on both sides. If this condition is combined with equation 1.6, one obtains $\rho_1 j_1^x = \rho_2 j_2^x$. Dividing both sides of the equation by the identical vertical current density $j_1^z = j_2^z$ one obtains $\rho_1 (j_1^x / j_1^z) = \rho_2 (j_2^x / j_2^z)$, which can be written as

$$\rho_1 \tan \theta_1 = \rho_2 \tan \theta_2 , \tag{3.3}$$

where θ_1 and θ_2 are the angles between the current lines and the boundary-normal in the two different media, respectively (see figure 3.4). If medium 2 is a perfect conductor ($\rho_2 = 0$ Ωm), it follows that $\tan \theta_1$ has to be 0 to fulfill equation 3.3. This means that $\theta_1 = 0°$, and the current lines intersect the boundary perpendicularly as postulated. For the perfect insulator ($\rho_2 = \infty$), the condition $\tan \theta_1 = \infty$ means that $\theta_1 = 90°$, and therefore the current lines are parallel to the surface. Both results are as predicted before.

In these two extreme cases (perfect conductor, perfect insulator), all values for tan θ_2 would fulfill equation 3.3, and no statement about the direction of any internal current flow can be made.

For the more realistic case of a finite resistivity contrast between the two media, the following can be concluded. If ρ_1 is smaller than ρ_2 (i.e., in figure 3.4 the current flows from a low resistivity medium at the top to a high resistivity medium below), then tan θ_1 must be bigger than tan θ_2 to fulfill equation 3.3. This means that current lines in the high resistivity medium are bent toward the normal; in other words, "higher ρ has smaller θ." With this consideration it becomes possible to draw diagrams of current distribution in the ground for features of arbitrary resistivity contrast. In a similar way plan diagrams of the horizontal current density at boundaries can be drawn (figure 3.5).

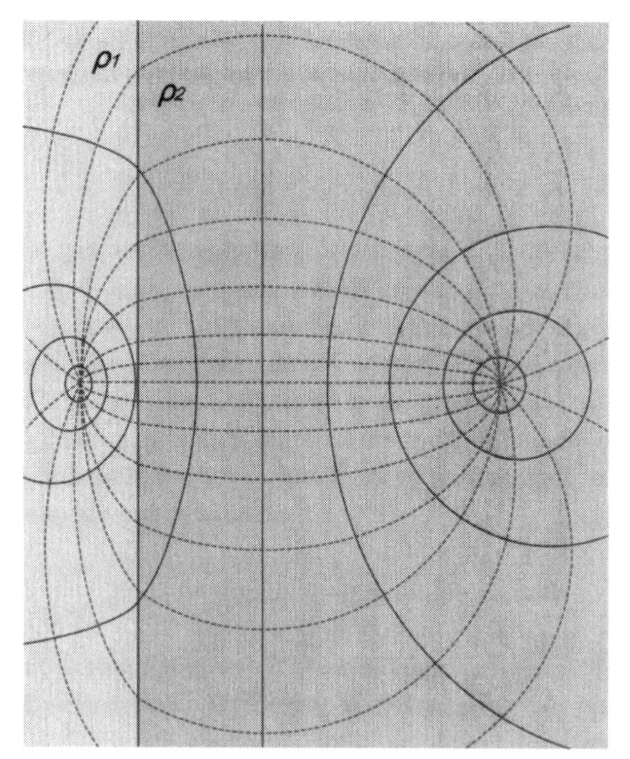

Figure 3.5. Horizontal view of refracted current lines and corresponding equipotential lines (broken and solid lines, respectively). The medium to the right has higher resistivity ($\rho_2 > \rho_1$) and current lines there are bent toward the normal.

3.4 Resistivity anomalies of a localized feature

Having investigated how current and potential lines in the ground are altered by the presence of features with resistivity contrast, it is important to consider the changes that are recorded in measurements on the ground surface. While the current may have to flow very differently near the buried features, the effects on the potential distribution at the surface can be very small (in figure 3.2, see the similarity of near surface equipotential lines to figure 2.8). These changes, compared to a homogeneous background, are usually referred to as "anomalies," and buried archaeological features in the ground create them. A geophysical interpretation of the anomalies attempts to associate them with the possible features in the ground: their resistivity contrast, size, and depth. To make such interpretations possible, it is necessary to know the anomalies that are produced by typical features.

While geological investigations mostly deal with extended layers of soil and rock, buried archaeological features are often discrete and localized. A typical example is a pit (in the form of either a "Grubenhaus"—a storage pit—or a post hole), but even ditches and foundations can be considered to be localized when viewed in cross section. The resistivity contrast of such features can be negative (e.g., for a pit that retains moisture, see section 1.4) or positive (e.g., a foundation stone or a pillar base). The simplest model for these features is a buried sphere of resistivity ρ_2 in a homogeneous medium that has a resistivity ρ_1. For such features anomalies can be calculated as shown by Lynam and Aspinall (Aspinall and Lynam 1970; Lynam 1970), and the following discussion will detail the anomalies that may be encountered for such localized buried features.

Considering that the overall shape of a measured anomaly will remain the same if all geometrical measurements are scaled by the same amount (e.g., by labeling graphs in meters instead of centimeters), it is useful to define all geometrical measurements relative to one characteristic dimension that acts as the reference scale. For the buried sphere this is the sphere's radius r (figure 3.6). So instead of the depth to the center of the sphere z, the relative depth $Z = z / r$ is used; instead of the electrode separation a, the relative electrode separation $A = a / r$ is used; and so on. This allows the discussion of a single case of, for example, a sphere at relative depth $Z = 1.4$ instead of many individual cases (such as $r = 0.5$ m and $z = 0.7$ m, or $r = 1$ m, $z = 1.4$ m, etc.). Given the way these relative

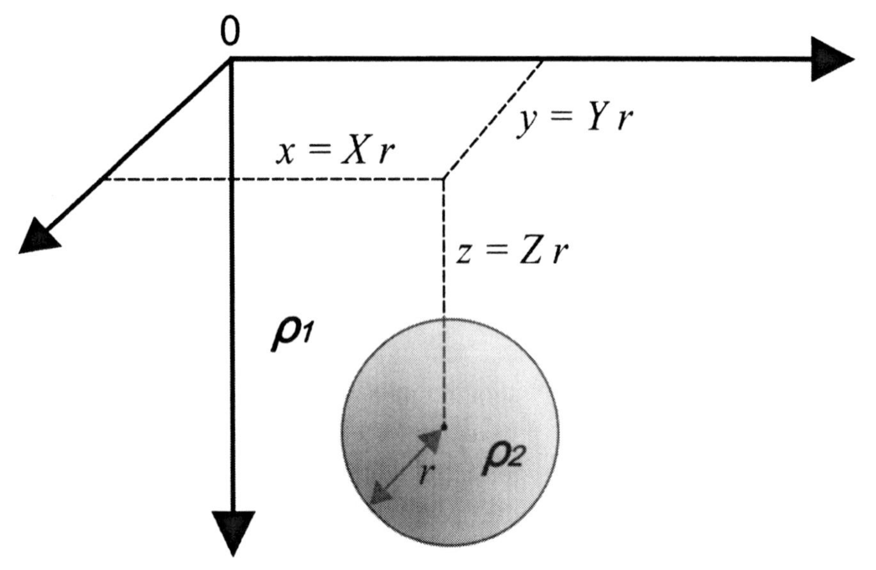

Figure 3.6. Coordinates of the buried sphere are expressed relative to its radius.

measures are formed, they are just numbers without units. For the following discussion the depth is always measured to the center of the sphere. For example, the top of a sphere with diameter of 0.5 m (r = 0.25 m) and a depth to the center of z = 0.35 m (thereby having Z = 1.4) would be 0.1 m below the ground surface. In the following discussion idealized spheres of either infinitely high or vanishingly low resistivity are being considered. The contrast factor introduced below can be used to convert these to real cases of finite resistivity values.

To make the discussion even more general the measured earth resistance is converted into apparent resistivity ρ_A (see section 3.1) and then related to the resistivity of the background material (ρ_1) away from the buried sphere. A "resistivity response" is then defined as the ratio ρ_A/ρ_1, which accordingly has a value of 1.0 for background measurements away from the feature and is larger or smaller than 1.0 for positive or negative anomalies, respectively.[3]

If a given electrode array is moved across a buried feature and readings are recorded at close intervals, the resulting data can be plotted as a

3. The anomalies commonly called "positive" and "negative" consequently have resistivity responses "larger than 1" and "smaller than 1."

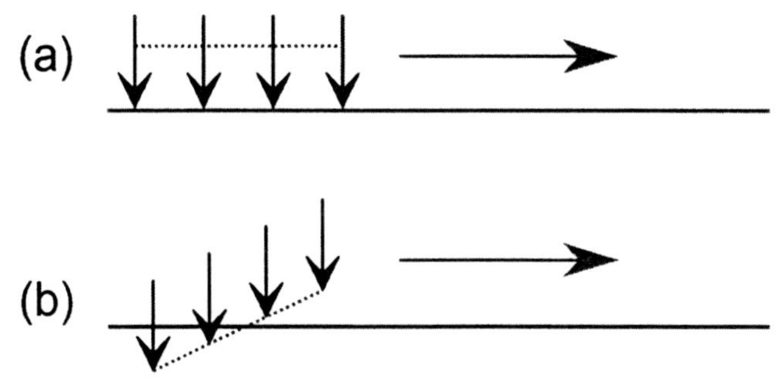

Figure 3.7. One can distinguish two directions of measurement traverses: (a) longitudinal; (b) transverse, also called "broadside."

continuous trace. As mentioned before, this is referred to as a traverse. If all four electrodes in an array are aligned in a line (e.g., in a Wenner array, figure 2.11), two cases can be distinguished. For a "longitudinal traverse" the electrodes are moved along their alignment (figure 3.7a), whereas in a "transverse traverse" (also called more intuitively "broadside traverse"), the line of electrode alignment is perpendicular to the direction of their movement (figure 3.7b). In the first case measurements may change as each individual electrode passes over the buried feature, creating complex anomalies. If used transversely instead, the electrodes pass over the feature nearly simultaneously, yielding a much simpler response.

The most important parameters influencing the shape of an anomaly are the relative depth Z of the buried sphere and the relative electrode separation A. It is therefore convenient to plot a number of transects in a diagram where one of these parameters is constant and the other is altered. However, it should be borne in mind that any combination of parameters is possible. This is similar to the two different ways the change of current density $j(z, L)$ can be looked at (see section 2.3).

3.4.1 Varying depth of feature

Figure 3.8 shows the anomalies measured with a longitudinal Wenner array ($A = 1$) when passing over perfectly insulating spheres at various depths. The relative distance X is measured between the center of the array and the center of the sphere. Assuming, as an example, that a buried pillar base has a diameter of 1 m and its top is 0.1 m below the surface, we can approximate it as an insulating sphere with a radius of $r = 0.5$ m. The

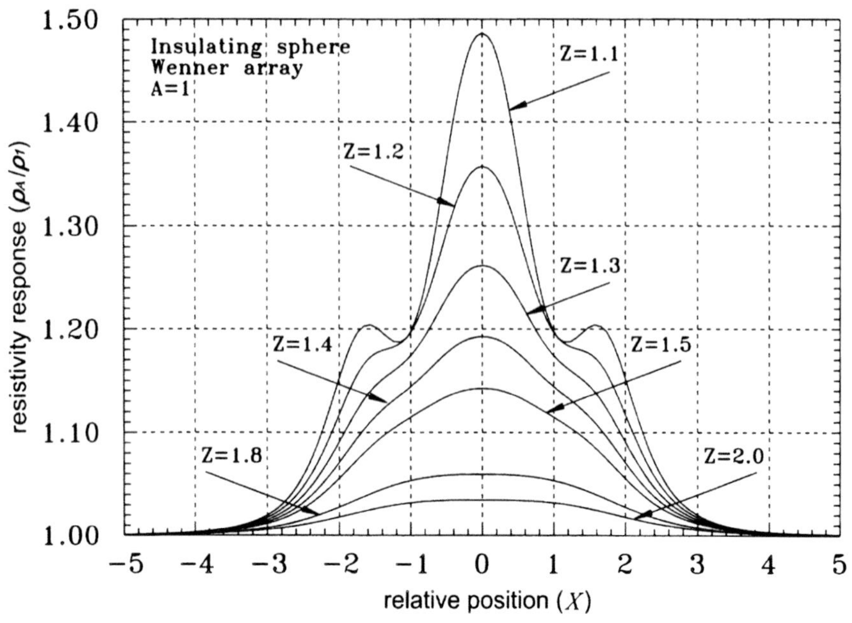

Figure 3.8. Theoretical resistivity response of a perfectly insulating buried sphere traversed by a Wenner array with relative electrode spacing $A = a/r = 1$**. The different curves are calculated for varying relative depth values** $Z = z/r$**.**

depth to its center is $z = 0.6$ m and hence $Z = 1.2$. A Wenner array with $A = 1$ then has an actual electrode separation of $a = 0.5$ m, which means that across all of its four electrodes it is 1.5 m wide.

For deeper spheres ($Z > 1.5$) the anomalies look broad, as expected. Since the sphere is a perfect insulator, it has a positive resistivity contrast resulting in an anomaly of the resistivity response that rises above 1. It varies smoothly and has its peak directly over the sphere's center. However, when the sphere is shallower (approximately for $Z < 1.5$), new phenomena occur. To the sides of the main peak, "shoulders" develop that become fairly pronounced for very shallow features. When interpreting such measurements it is important to realize that they are not caused by three successive features, but by one single sphere at shallow depth that has this characteristic signature when measured with a Wenner array. For an oversimplified explanation of this behavior one can consider that at very shallow depth of a sphere the three pairs of electrodes in the Wenner

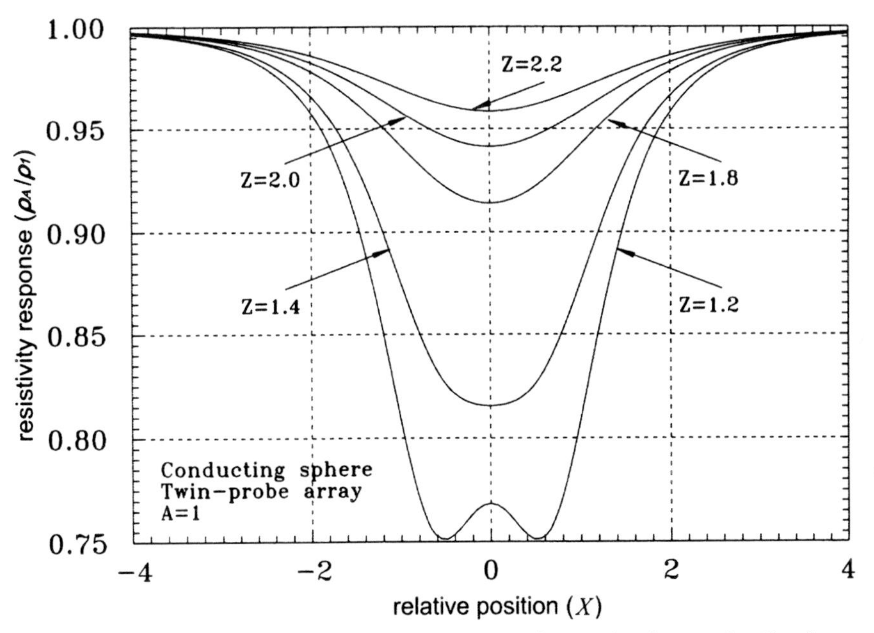

Figure 3.9. Theoretical resistivity response of a perfectly conducting buried sphere traversed by a twin-probe array with relative electrode spacing $A = 1$. The curves are plotted for different depths Z.

array pass over the sphere one after the other, nearly as if they were independent, leading to three "peaks."

Similar observations can be made for measurements with a twin-probe array. For the perfect conductor in figure 3.9 the separation of the two mobile electrodes was chosen to be identical to the radius of the sphere ($A = 1$), and the electrode array is then half as wide as the buried feature. In this case the results are fairly simple. The negative resistivity contrast leads to a single trough in the traverse. For very shallow spheres a slight peak appears in its center.

However, if the twin-probe array is as wide as the feature ($A = 2$, figure 3.10), results become more complicated. Now even at moderate depth ($Z = 1.8$) the peak in the middle of the trough is fairly pronounced, and for $Z > 1.2$ the resistivity response becomes even larger than 1.0, which would normally be expected only for an insulating feature.

To illustrate how this information can help to interpret survey results, figure 3.11 shows earth resistance data from a suspected graveyard near

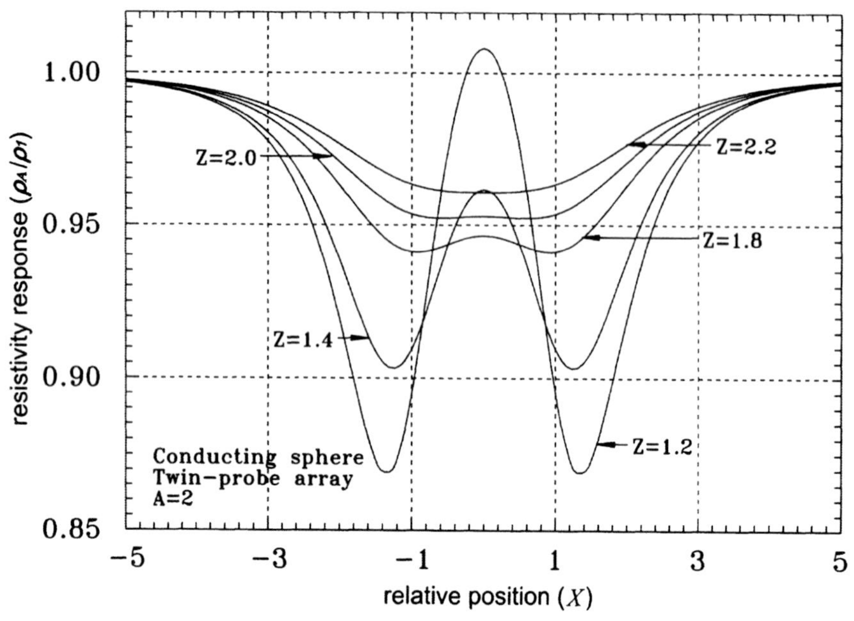

Figure 3.10. Theoretical resistivity response of a perfectly conducting buried sphere traversed by a twin-probe array with relative electrode spacing A = 2. The curves are plotted for different depths Z.

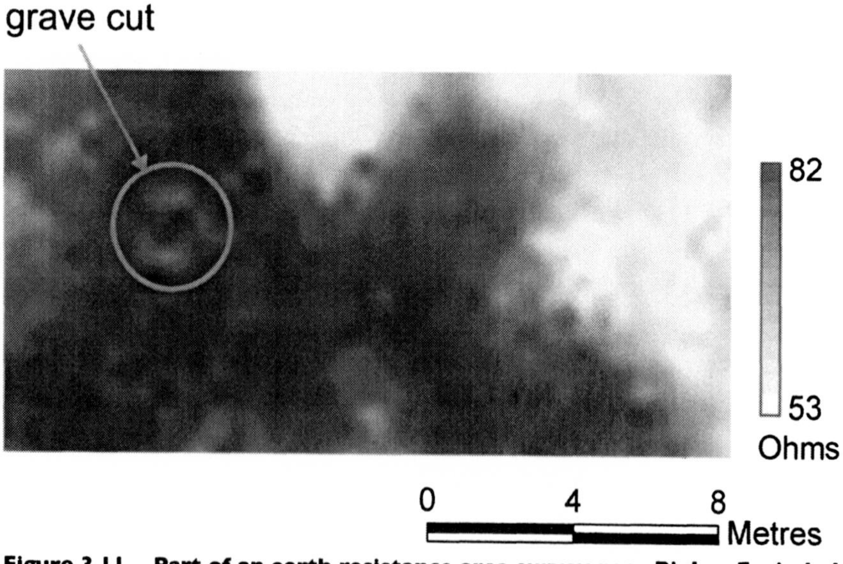

Figure 3.11. Part of an earth resistance area survey near Ripley. Encircled is a triplet anomaly (low-high-low, from north to south) that is typical for a single narrow buried feature at shallow depth, most likely a grave cut.

Ripley (North Yorkshire, United Kingdom). Although all anomalies were already visible in the raw data, they were interpolated for a more pleasing display. The marked anomalies show a high earth resistance peak (dark gray) flanked by two smaller low resistance anomalies (light gray), one north and one south. This is exactly as predicted for a conducting object (i.e., the grave cut) of about the size of the electrode array ($A \approx 2$). In this survey the electrode separation was $a = 0.5$ m, and one can therefore estimate a feature radius of $r = a/A \approx 0.25$ m; that is, a feature width of 0.5 m. The results are similar to those expected for a shallow depth of $Z \approx 1.2$. Consequently, the depth to the center of the grave is approximately 0.3 m. Knowing the shapes of shallow twin-probe anomalies allows avoiding a misinterpretation of the results as being caused by a central high-resistive feature flanked by two smaller conductive features. The measured triplet anomaly is caused by only one single conducting feature! To interpret the results from this survey correctly, it was necessary to know the geophysical characteristics of twin-probe data as well as having the archaeological information that the site may contain shallow burials.

A similar effect is seen in figure 3.12 for small patches of very wet grass in a field that was heavily disturbed and trampled by horses, and where the survey was undertaken after heavy rain. In this twin-probe survey north of the Roman fort at Slack (West Yorkshire, United Kingdom), again with $a = 0.5$ m, several high earth resistance anomalies are each surrounded by a pair of low anomalies, one north, one south. Each of these triplets is again interpreted as the result of one single conducting feature with $A \approx 3$, leading to an estimate for the diameter of the wet grass patches of ca. 0.33 m.

From the discussion so far it has become clear that the *shape* of an earth resistance anomaly is characteristic for the electrode array used and the geometrical parameters of the buried features. But what about the strength of the anomaly? Unfortunately, the peak height depends not only on the depth of a feature (as indicated in figures 3.8–3.10) but also on its resistivity contrast. As the contrast varies from positive through zero to negative, the resistivity response changes from above 1, through 1, to below 1. For spheres with a finite resistivity the "contrast factor" is

$$k' = \frac{\rho_2 - \rho_1}{2\rho_2 + \rho_1} \tag{3.4}$$

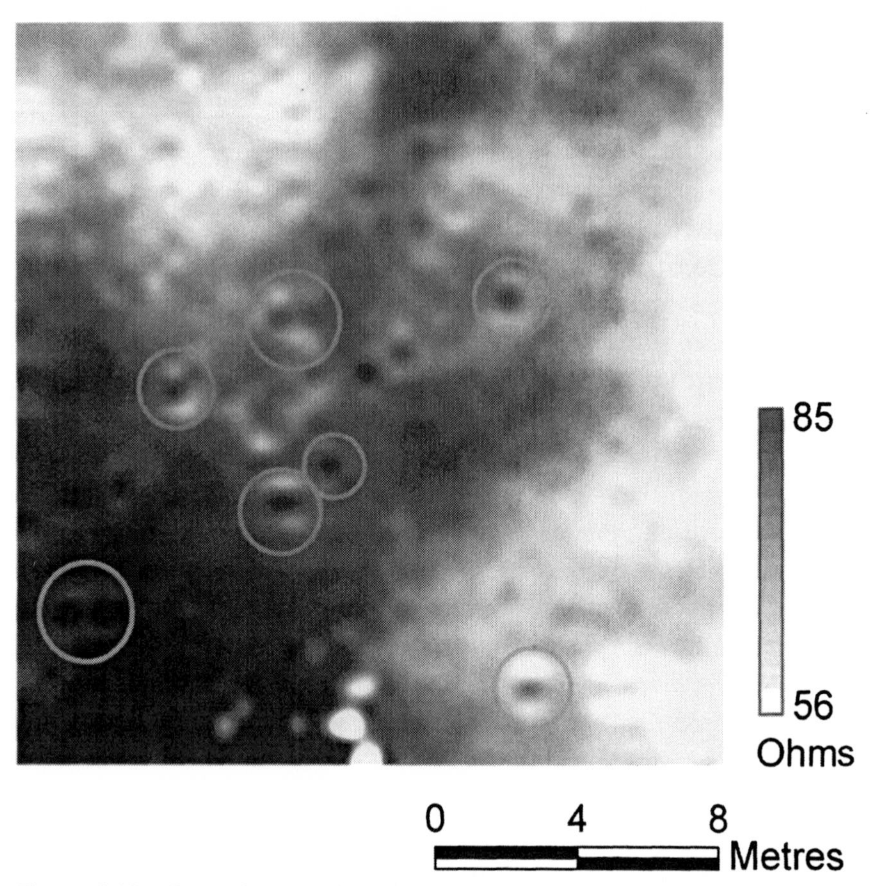

Figure 3.12. Part of an earth resistance area survey at the Roman fort at Slack. The encircled triplet anomalies (low-high-low, from north to south) are each caused by one small and shallow conducting feature, probably a patch of very wet grass in a field trampled by horses.

and has to be applied to the perfect conductor's response curve[4] to obtain the correct resistivity values for a buried feature. If there is no resistivity contrast ($\rho_2 = \rho_1$), then $k' = 0$ and no anomaly is created. For a perfect insulator ($\rho_2 = \infty$, positive resistivity contrast), the expression yields $k' = +0.5$; for a perfect conductor ($\rho_2 = 0\ \Omega$m, negative resistivity contrast), $k' = -1$. Thus the trough of the conducting anomaly is twice as deep as the insulator's peak. This

4. Since $k' = -1$ for the perfect conductor (negative contrast), the response curve of the perfect conductor (which already has a dip) has to be multiplied with $-k'$ to yield the response curve for any finite resistivity of ρ_2.

asymmetry between conducting and insulating features is a result of the measurements being made from an insulating medium (air) into the conducting ground (Aspinall and Lynam 1970; Lynam 1970) and manifests itself in the factor of 2 in the denominator of equation 3.4. This asymmetry is also the reason why this somewhat unusual contrast factor was labeled *k'* and not *k*.

3.4.2 Varying electrode separation

Having considered the anomalies produced by spheres at various depths it will now be investigated what effect the electrode separation has on the shape of the resistivity responses. In figure 3.13 the twin-probe anomalies of a very shallow insulating sphere (Z = 1.1) are shown for relative electrode separations ranging from A = 0.5 to 3.0. It can be seen clearly that narrow twin-probe arrays, compared to the diameter of the sphere (A < 0.7), produce simple results, while with wider separation the complexity of anomalies increases markedly.

Similarly for a deeper sphere (Z = 1.5) and a Wenner array (figure 3.14), the anomalies attain a rather complicated shape if the relative

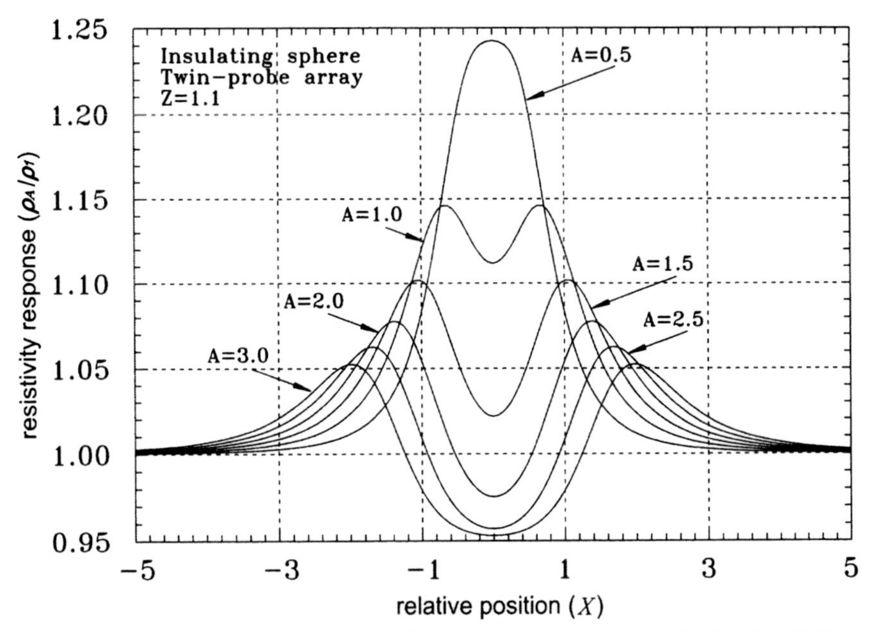

Figure 3.13. Theoretical resistivity response of a perfectly insulating buried sphere at relative depth $Z = z/r$ = 1.1. The curves are calculated for traverses with twin-probe arrays that have different relative electrode spacing $A = a/r$.

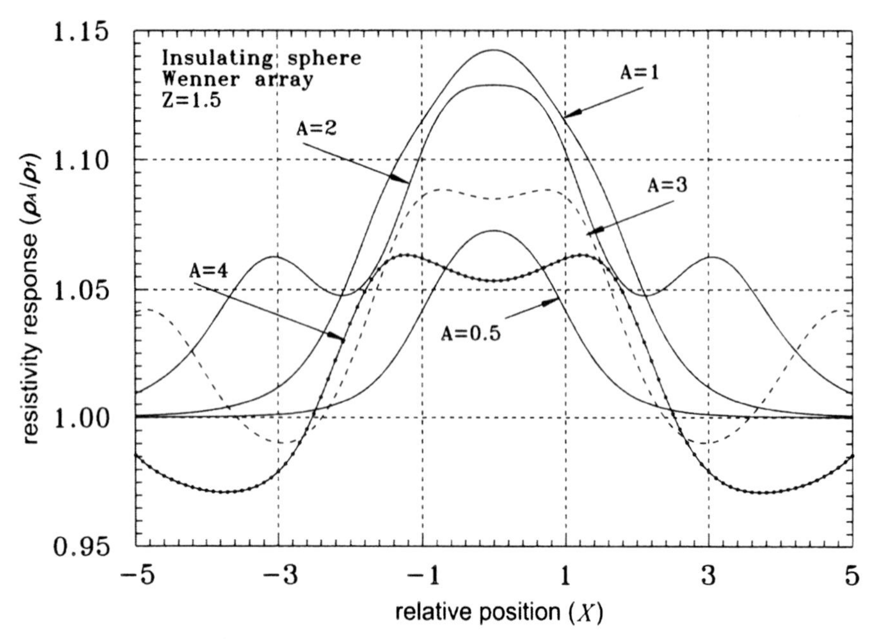

Figure 3.14. Theoretical resistivity response of a perfectly insulating buried sphere at relative depth Z = 1.5. The curves are calculated for traverses with Wenner arrays that have different relative electrode spacing A.

electrode separation rises beyond approximately $A = 2.5$. At that separation the overall width of the array, over all four electrodes, is 3.75 times the diameter of the feature.

In addition, it can be seen that the peak height initially rises with increasing electrode separation, but then steadily decreases for $A > 1$. This behavior can also be seen for spheres at other depths and is shown in the graph of figure 3.15, where the strength of the response in the center of the Wenner array is plotted against the electrode separation. For each depth of the sphere the maximum resistivity response is encountered at a different electrode separation.

3.4.3 Vertical Electrical Sounding

Over a buried feature the dependence of the apparent resistivity on the electrode separation[5] can be exploited to derive additional information

5. For the homogeneous ground the apparent resistivity is independent of the electrode separation.

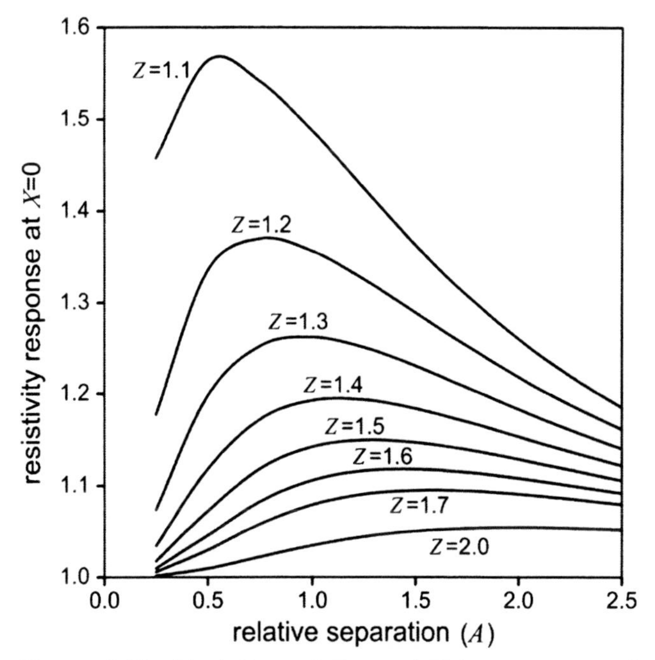

Figure 3.15. Variation of the resistivity response of a Wenner array with increasing relative electrode separation *A*. The array is kept centered over a buried sphere (X = 0). The curves represent spheres at different relative depth *Z*.

about that feature. In a Vertical Electrical Sounding (VES) the electrode array is gradually expanded about its central position (figure 3.16), and earth resistance is measured for each separation. These values are then converted to apparent resistivity or to a resistivity response (i.e., the apparent resistivity relative to the background value away from the feature) and plotted against the electrode separation. In this way a characteristic "sounding curve" is obtained. It can be compared to calculated sounding curves, like those in figure 3.15, and an estimate of the depth of the buried features can be obtained. VES measurements will be discussed in more detail in section 3.5.1 for the layered earth.

3.5 The layered earth

As explained above, features encountered in archaeological prospecting are often localized and can be approximated by a buried sphere. In some

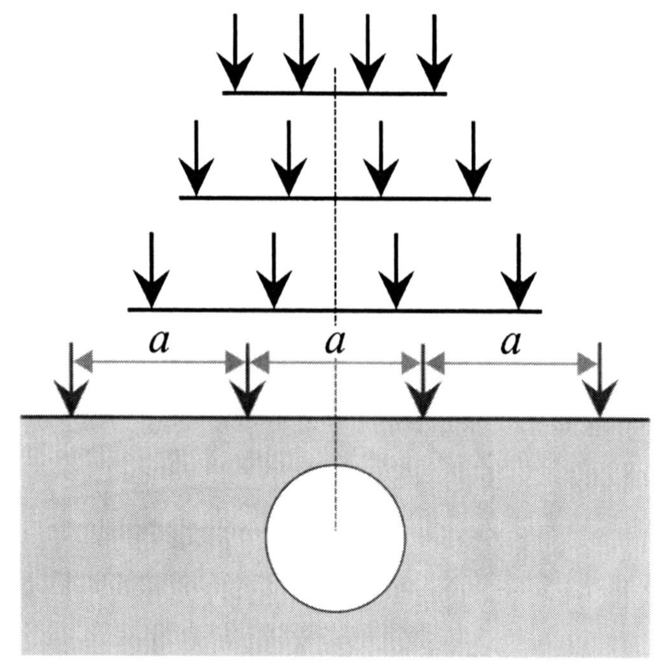

Figure 3.16. For a Vertical Electrical Sounding, a Wenner array can be expanded over the center of a buried sphere.

instances, however, it is helpful to use the model of a layered earth that was initially developed for geological investigations. Typical archaeological features that may be described by this model are floor surfaces, mosaics, fishponds, or other extended features that form horizontal boundaries between layers. For these cases the geophysical investigation is mainly concerned with the depth of these boundaries and the resistivities of the individual layers. VES measurements can help to obtain these parameters.

3.5.1 Two layers

In the "two layer model" a horizontal top layer of thickness z and resistivity ρ_1 overlays a layer of resistivity ρ_2 that is extending further downward without lower boundary (figure 3.17a). For the case where the top layer has lower resistivity ($\rho_1 < \rho_2$) (e.g., moist silty soil in a former fishpond, with a hard impermeable base, or agricultural soil over a compacted floor horizon), it is plausible that the current injected will "prefer" to flow

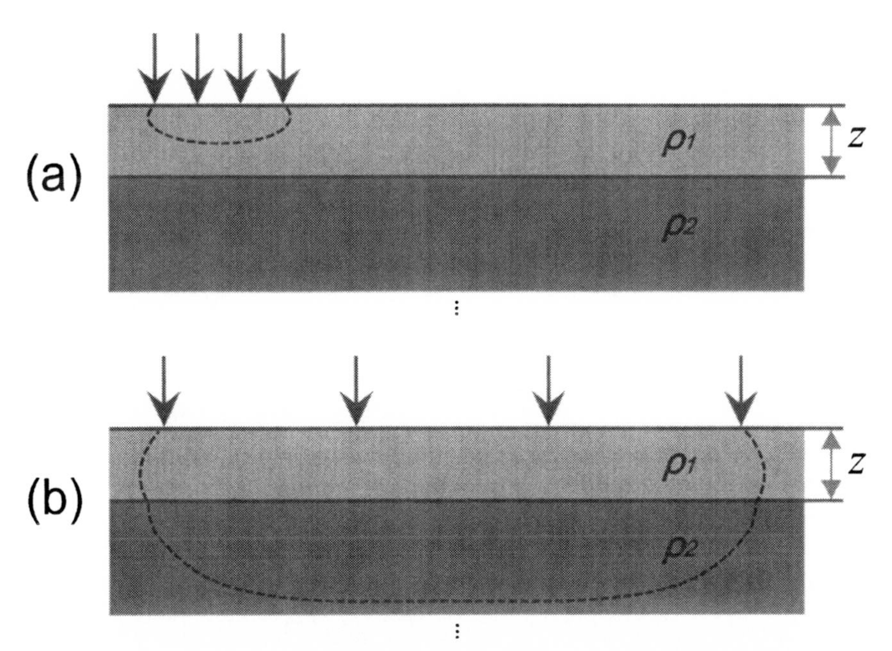

Figure 3.17. To describe current flow through the layered ground the top layer has a thickness of *z* while the bottom layer is unbounded downward. Current lines are drawn for the case where the top layer has lower resistivity: (a) for very close electrode separation nearly all current flows through the top layer; (b) for very wide electrode separation nearly all current flows through the bottom layer.

through the low resistivity top layer and only choose the higher resistivity bottom layer when it really has to. The current lines will be refracted at the boundary as discussed in section 3.3.

For very close electrode separation (see figure 3.17a), nearly all current will flow in the top layer and the apparent resistivity of the measurement will therefore be very close to ρ_1. For the other extreme case of a very wide electrode separation (figure 3.17b), the top layer will carry only a small fraction of the total current and the apparent resistivity will be very similar to ρ_2, although it will always be influenced slightly by the conducting top layer. Figure 3.18 (solid line) visualizes this relationship. This so-called sounding curve is plotted on a double logarithmic scale that reflects the asymmetry between these two extreme cases.

For the opposite case of a high resistivity top layer ($\rho_1 > \rho_2$) (e.g., a hard floor surface over a modestly wet subsoil like clay), the situation for

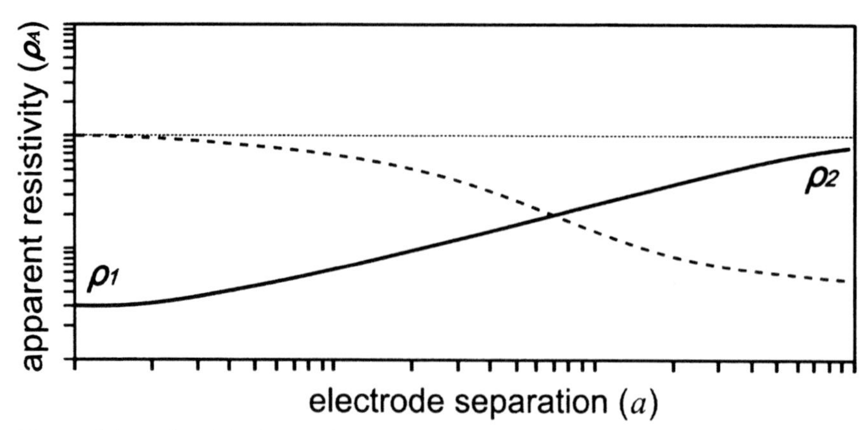

Figure 3.18. Double logarithmic sounding curves can be calculated for the layered earth: solid line for a conducting top layer ($\rho_1 < \rho_2$); broken line for a resistive top layer ($\rho_1 > \rho_2$).

close electrode separation is similar insofar as all current will remain in the top layer and $\rho_A \approx \rho_1$. For the wider electrode separation, however, the current will flow in the lower layer much more readily than in the previously discussed case due to its low resistivity. The corresponding sounding curve in figure 3.18 (broken line) is therefore not symmetric to the previous case.

The exact shape of the sounding curve can be calculated for various electrode arrays (Keller and Frischknecht 1966) and is often specified in a form of

$$\frac{\rho_A}{\rho_1} = \mathrm{f}_1\left(\frac{a}{z}, \frac{\rho_2}{\rho_1}\right) \text{ or } \frac{\rho_A}{\rho_1} = \mathrm{f}_2\left(\frac{a}{z}, k\right), \tag{3.5}$$

where

$$k = \frac{\rho_2 - \rho_1}{\rho_2 + \rho_1} \tag{3.6}$$

is the contrast factor for the two-layer model,[6] a is the characteristic separation of the electrode array, and f_1 and f_2 are the functions that relate

6. Note that this definition is different from the contrast factor for the buried sphere due to the different symmetry of the model.

these parameters. The notation above simply means that the relationship is expressed through a two-parameter function in which the first parameter is the value *a/z*. In the simple case in which the bottom layer is a perfect insulator (e.g., bedrock, hence $\rho_2/\rho_1 = \infty$, i.e., $k = +1$) and the electrode separation is large compared to the top layer's thickness, it can be approximated for a Wenner array as

$$\frac{\rho_A}{\rho_1} = f_1\left(\frac{a}{z}, \infty\right) = 1.38\frac{a}{z} \tag{3.7}$$

(Keller and Frischknecht 1966). This equation represents a straight line through the origin (figure 3.19).

In measurements where it can be assumed that the bottom layer is a good insulator, this relationship can be used to calculate the unknown parameters ρ_1 and z from a Wenner sounding curve (see figure 3.19). For a very close electrode separation the apparent resistivity is an approximation

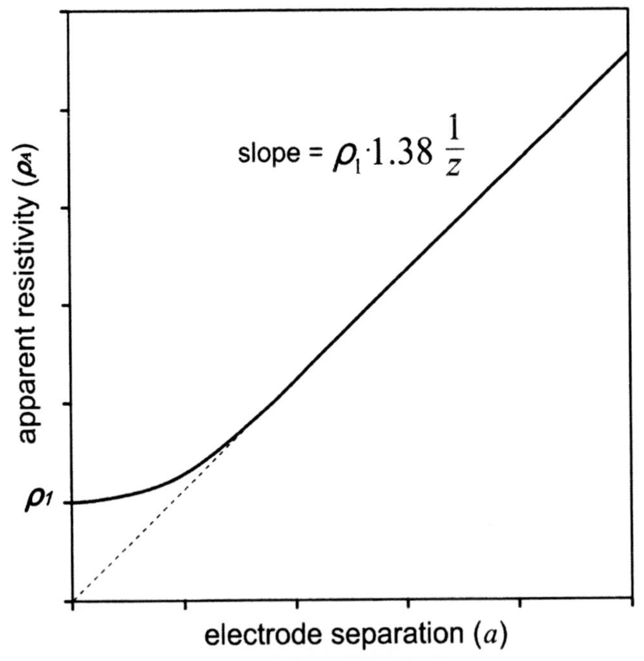

Figure 3.19. Dependence of measured apparent resistivity ρ_A on the Wenner electrode separation *a* for a perfectly insulating bottom layer.

for the top layer's resistivity since only a small volume of soil is involved: $\rho_1 = \rho_A(a \approx 0)$ (see above). The slope of the sounding curve derived from a straight-line fit at large electrode separations can then be used to calculate the depth to the interface z from equation 3.7, since $\rho_A = (\rho_1 \cdot 1.38 \cdot z^{-1})\, a$ and the fitted slope must be equal to the value in the bracket, allowing the calculation of z (ρ_1 was calculated in the previous step).

For more general cases graphical methods are often used to obtain the parameters of the layered earth. Sets of "master curves" have been published (Keller and Frischknecht 1966) that can be used to match measurement results and find the required information. Figure 3.20 shows master curves for Wenner sounding measurements over a two-layered earth. As previously discussed, results for low and high resistivity top layers are different, and this is clearly shown by their different shape ($\rho_1/\rho_2 = +\infty$ and $-\infty$). In addition, these curves are completely different from those obtained above for the localized feature (see figure 3.15), even though the measurement procedure of expanding an array over a fixed location is exactly the same.

Since the sounding curves are drawn in relative coordinates (a/z and ρ_A/ρ_1) on a double logarithmic scale, the measured sounding results in a diagram of ρ_A against a plotted on a transparent sheet can simply be slid over them to find the matching curve. This yields the information about the relative layer resistivities ρ_1/ρ_2. Where $a/z = 1$ and $\rho_A/\rho_1 = 1$ on the master curve, the values for z and ρ_1 can easily be read off the measurement results. Sounding curves for other arrays have been published and can be found in the relevant literature (Keller and Frischknecht 1966).

Figure 3.21a shows a Wenner Vertical Electrical Sounding (VES) along a Roman ditch, undertaken by Peter Cott at Caistor, Norwich, United Kingdom (Cott 1997). An area survey of the ditch showed that the fill was fairly homogeneous. This meant that expanding the electrode array along the ditch's long axis (figure 3.21b) sampled for small electrode separations the nearly homogeneous soil of the fill, with a resistivity similar to the apparent resistivity of the 0.5 m electrode separation ($\rho_1 = 427\ \Omega$m). Only when the main current density reached the bottom of the ditch did the injected current start to flow through the chalk subsoil into which the ditch was cut. This situation was approximated by a two-layer model. By selecting different trial values for the thickness of the top layer z (i.e., the approximate depth of the ditch) and plotting the measurements as response

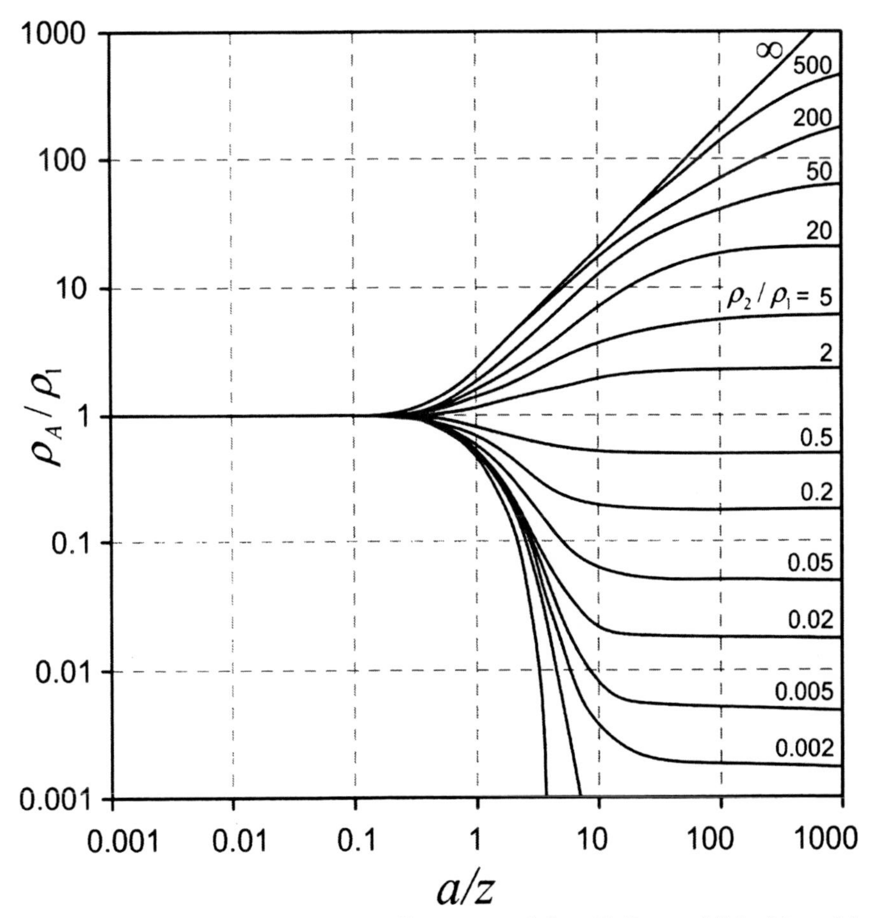

Figure 3.20. Set of Wenner sounding curves (after Keller and Frischknecht [1966]).

ρ_A/ρ_1 against the relative separation *a/z*, a good fit with the Wenner sounding curve of ρ_2/ρ_1 = 1/5 was achieved for the value of z = 3.0 m. This thickness of the top layer was therefore considered to be the approximate depth to the bottom of the ditch, and the matching resistivity ratio allowed the calculation of the resistivity of the wetter subsoil as ρ_2 = 85 Ωm.

A control measurement was recorded parallel to the ditch. Its resistivity response (broken line in figure 3.21a) started with a higher value (513 Ωm), since the chalk subsoil drained this area more effectively than the fill of the ditch. The contrast between this drier top layer and the wetter subsoil, the same that is underneath the ditch (85 Ωm, see above), was

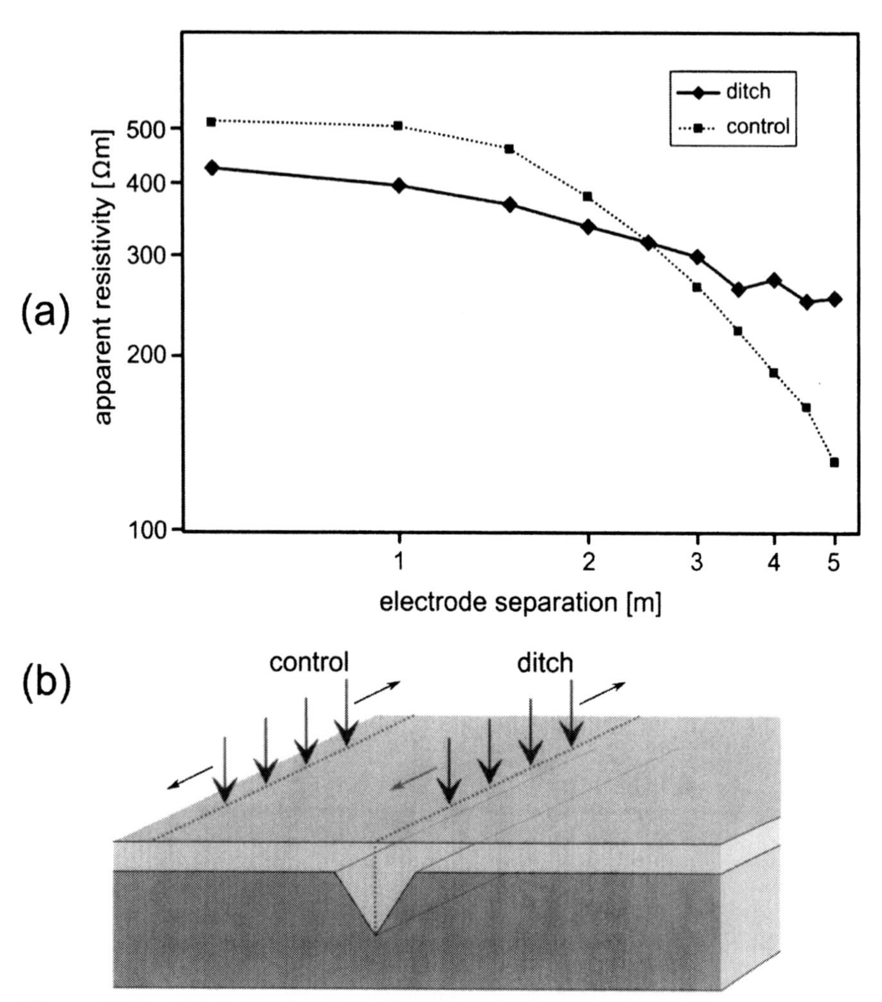

Figure 3.21. Wenner Vertical Electrical Sounding was measured along a Roman ditch: (a) results from the ditch and adjacent control line; (b) layout of the survey lines.

hence bigger than there (ca. 1/6), resulting in the steeper sounding curve of the control data. However, when overlaying these measurements on the Wenner sounding curves for the two-layer model the match was poor, and the best estimate was for z = 2.4 m and ρ_2 / ρ_1 = 1/20. This model therefore appeared inadequate outside of the ditch, and it was concluded that instead of two distinct layers the soil away from the ditch showed a more gradual change of resistivity with depth.

3.5.2 Computing the layers' resistivities

As discussed before, the simple case of a two-layer model can be described by three parameters alone: ρ_1, ρ_2, and z. Using the methodology of matching data with existing sounding curves, as explained above, it is possible to derive these parameters from the measurements. However, if more than two layers are involved, the number of possible cases increases rapidly, and producing master curves for all of them becomes nearly impossible. It is therefore necessary to calculate parameters with appropriate software, and various algorithms have been published. The method suggested by Zohdy (1989) and Loke and Barker (1995a) is now frequently used. First, the ground is subdivided into as many layers as there are individual measurement points in the sounding curve (figures 3.22a and b). To initialize the computer algorithm, the bottom of each layer is assumed to lie at a depth equivalent to the electrode separation used for the measurement, and

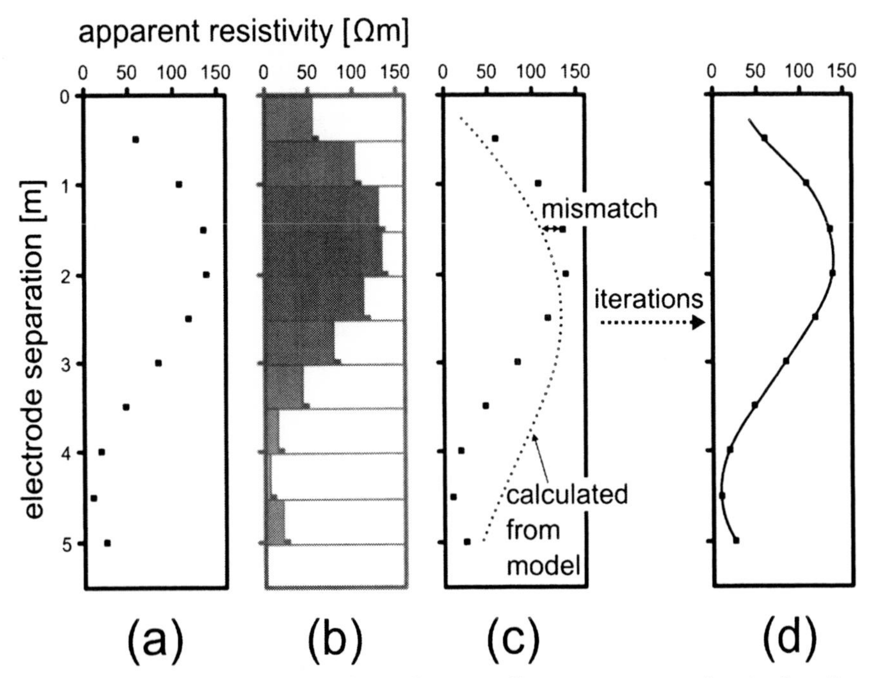

Figure 3.22. Through inversion of a sounding curve an estimate for the ground resistivities can be obtained. The resistivity distribution in the ground is computed by iteratively adjusting a model until the apparent resistivity calculated from the model matches the measurements. See text for details.

its resistivity is provisionally set to the respective apparent resistivity (see figure 3.22b). This leads to an initial model for the layered earth and its resistivity values. The software then calculates the resulting sounding curve that would be measured over such a set of layers (the "forward modeling," figure 3.22c). These results will be different from the actual measurements, as this first guess was only a crude approximation, and the software then adjusts the model's resistivities to reduce this error. The sounding curve is calculated again, and resistivities are further adjusted. That process is iterated until a set of resistivity values has been found that leads to a satisfactory match between calculated and measured sounding curves (figure 3.22d). This process of finding soil parameters that match the geophysical measurements is called inversion. The algorithm is fast and produces reliable results even where the soil's resistivity varies smoothly with depth instead of showing sharp resistivity boundaries. Unfortunately, the solution is not unique, since a variety of combinations of depth and resistivity values may lead to virtually the same sounding curve, especially if there are slight statistical variations in the measurements ("noise"). This problem of nonunique solutions is inherent to all geophysical inversions and stems from the problem of estimating parameters of the full three-dimensional ground from measurements from the two-dimensional surface.

3.6 Other resistivity distributions

While closed mathematical solutions are available for the buried sphere and the layered earth, most other features require computational modeling to derive predictions for earth resistance anomalies ("forward modeling," see above)—for example, with software such as Res2DMod. These calculations may provide further insights into the shape of an anomaly if the buried feature is well known. However, for the interpretation of measured earth resistance data over unknown features these models provide only somewhat more insight than what can be deduced from the two extreme cases discussed above. The main conclusion derived from the buried sphere is that even simple features can create complex multipeak anomalies, and the delineation of features from a map of these anomalies has to be undertaken very carefully. Electrode configuration, and possible feature depth and size, must be taken into account. The simple single-peak anomaly that would be desirable (Witten 2006; see figure 13.9) is not always what the data show. The case of the layered earth showed that apparent resistivity and soil resistivity depend on each other but are not identical for most situations. Earth resistance area investigations clearly are not "resistivity surveys."

CHAPTER FOUR
RESISTIVITY IMAGING

4.1 Depth of investigation

In the previous discussion of Vertical Electrical Sounding (VES) it was explained how expanding electrode separations are linked to increasing depths of investigation. Figure 4.1 illustrates that the further apart the electrodes are, the deeper the area is that influences the measured earth resistance. While this qualitative insight is useful, it has to be remembered that the measured earth resistance is influenced by *all* material in the ground—to a larger or smaller extent. This can be seen from the change of current density with depth (see figure 2.7) as discussed in section 2.3. At no point does the current density fully disappear; it decreases continuously with depth, and even features at great depth can have an influence on the measured earth resistance if they exhibit a large enough resistivity

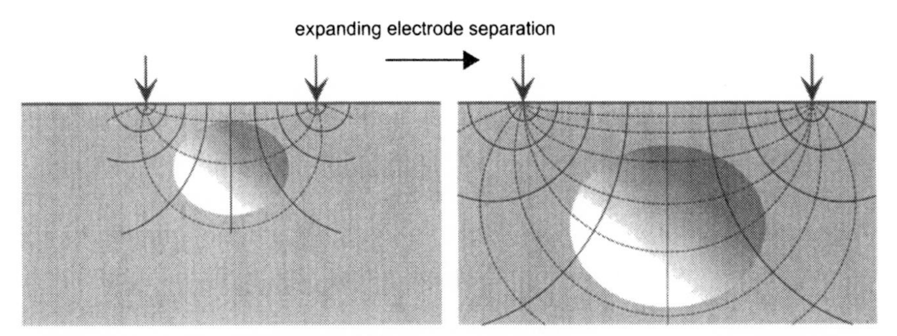

Figure 4.1. When electrodes are further apart the measured earth resistance is influenced by features that are deeper.

contrast. Nevertheless, many attempts have been made to define a single "depth of investigation" (DoI) that could be used to assign a characteristic depth to a given electrode separation (Roy and Apparao 1971; Barker 1989; Apparao et al. 1992; Apparao and Srinivas 1995).

When discussing current density in the homogenous earth (section 2.3), it was shown that an expanding electrode array produces the largest current density at a given depth z if the separation of current electrodes is $L = \sqrt{2}z$. This is therefore one possible candidate for a DoI. Alternatively, one could look for the strongest change of current density with depth, which can be calculated to be[1] $L = 4\ z$. At this depth the current density shows the maximum sensitivity and can therefore be influenced most strongly by a resistivity contrast. One can also consider that the measured earth resistance is influenced by all the material above the DoI through which the current has to flow. It is therefore sensible to integrate the current density downward to find the depth up to which the majority of the total current density flows (e.g., 90 percent). This depth can then be used as another feasible estimate for the DoI. In many cases, however, an empirical factor is chosen that provides "best looking" results, when compared visually with true resistivity data. Typical values are 1/2 or 1/3 of the electrode separation a ([C_1P_1] for a Wenner or twin-probe array) (Griffiths and Barker 1994). Using this estimate the empirical relationship

$$DoI = a/2 \tag{4.1}$$

has proven appropriate for many archaeological cases.

4.2 Pseudosections

In a "pseudosection," the above-described principle is used to display results from measurements where the electrode separation of a selected array is systematically expanded and the array is then moved forward along a line to cover a whole profile. Such measurements are referred to as basic Electrical Resistivity Imaging (ERI). For such a basic ERI the results can be visualized by assigning the apparent resistivity of each measurement

1. By forming the second derivative of equation 3.8.

to the depth of investigation (DoI) underneath the center of the relevant electrodes. This results in a vertical apparent resistivity display that is related to the actual resistivities of the ground and is called a pseudosection. The name *pseudo*-section is a reminder that this procedure wrongly assumes a direct relationship between an earth resistance measurement and the resistivity at one single depth. However, as examples have shown (see below), it is nevertheless a useful, objective, and very simple display to gain information about the depth distribution of soil resistivities.

Depending on the method of data collection, two types of basic ERI profiles can be distinguished. The "fixed center" ERI consists of a sequence of VES, where the chosen electrode array (e.g., Wenner) is expanded about its central point (see figure 3.16). The electrodes can be expanded either longitudinally along the line of the ERI profile or transversely (i.e., perpendicular to it) (figure 4.2). The disadvantage of this method is that all electrodes of the array have to be moved for each expansion of the electrode separation. Alternatively, using the "shifting center" arrangement, fewer electrodes have to be moved for each expansion step by keeping the first electrode in place until the full expansion of the array is reached (figure 4.3d), and only the remaining electrodes are moved. For a Wenner array, for example, only three electrodes have to be moved in this case (C_1 is kept fixed). As a result, the center of the array shifts for increasing electrode separations, and in a pseudosection plot this results in a

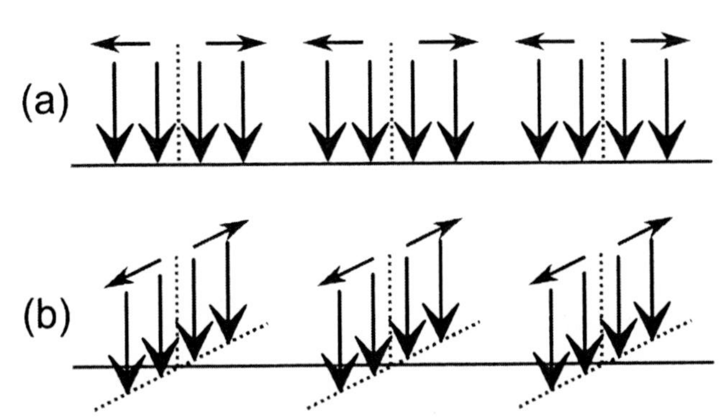

Figure 4.2. In a fixed center ERI the electrodes are gradually expanded about their center at each position: (a) longitudinal expansion; (b) transverse expansion perpendicular to the line of the ERI.

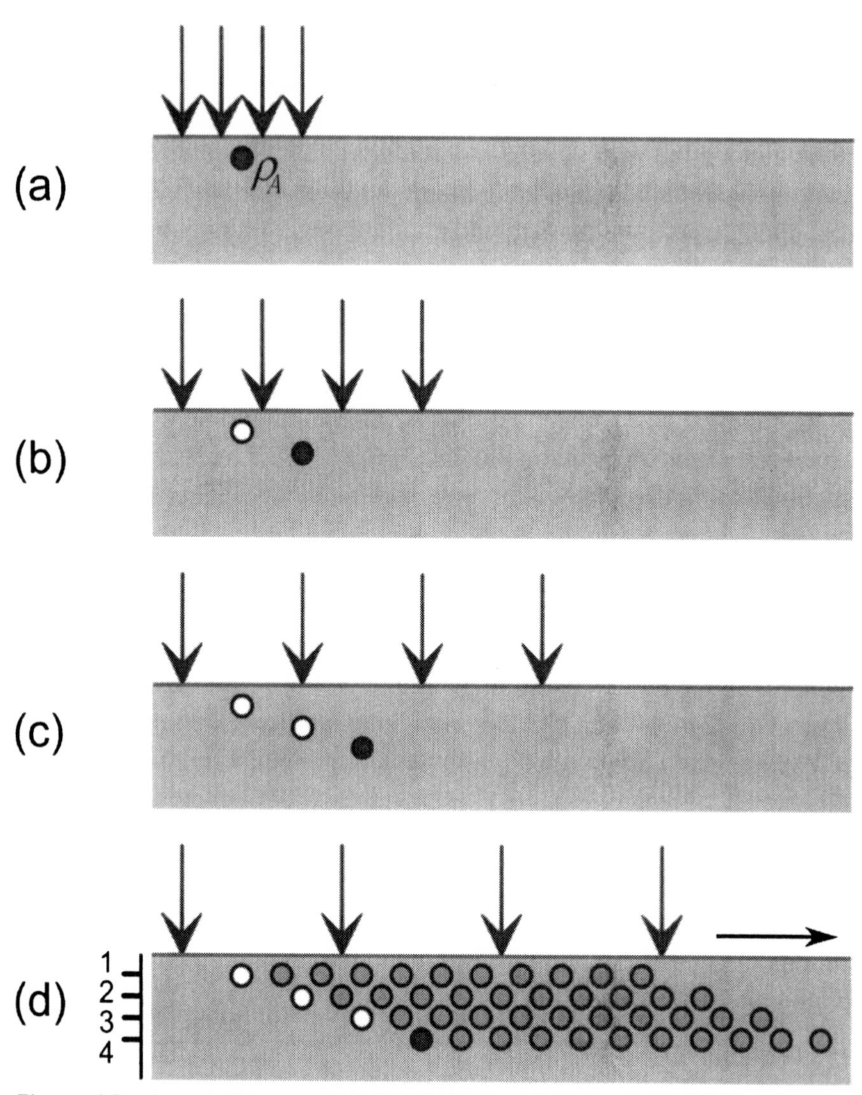

Figure 4.3. In a shifting center ERI the first electrode is kept in a fixed position and the remaining electrodes are gradually expanded; (a) to (d). Then the first electrode is moved and the array is expanded again, thereby building up a pseudosection of trapezoidal shape (gray dots). For each electrode arrangement the apparent resistivity is plotted underneath the center of the array at an appropriate depth (black dots).

non-rectangular mesh of data (see figure 4.3d) with a trapezoidal shape of the overall pseudosection. For this acquisition method, only longitudinal electrode expansions are possible.

As an example, figure 4.4 shows the pseudosection results of a Wenner ERI profile over two ditches in a shallow water tank. The model tank was constructed from acrylic sheets and filled with tap water. The resulting pronounced and sharp resistivity contrast helped to produce a pseudosection that closely resembles the investigated structure. For this pseudosection the depth of investigation was chosen as 0.5 *a* (equation 4.1), since this produced results most closely resembling the actual model.

It is also possible to use a twin-probe array for basic ERI profiles. In this case the data acquisition is only performed with the two mobile electrodes (C_1 and P_1), and the two remote electrodes remain fixed at a distance. When using a twin-probe configuration for a "shifting center" ERI, only one electrode has to be moved when expanding the array, which greatly reduces the required effort. If an earth resistance instrument is used that automatically triggers and records measurements upon insertion of the current electrode in the ground (e.g., RM15 from Geoscan Research), the ERI can easily be recorded by a single person (figure 4.5) and then displayed as a pseudosection.

Although it is possible to insert the electrodes manually in the right location for each new earth resistance measurement, this can be a laborious procedure, and therefore automated electrode arrays have been

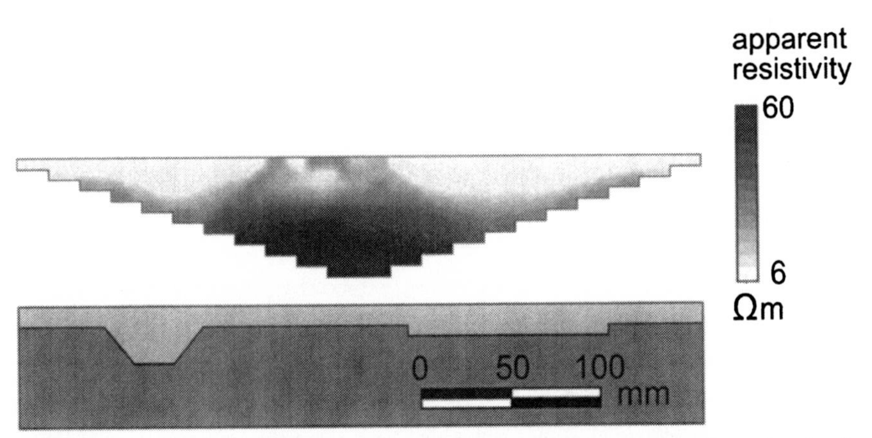

Figure 4.4. The pseudosection of a Wenner ERI from an acrylic water tank with a deep and a shallow ditch provides a good approximation of the actual features.

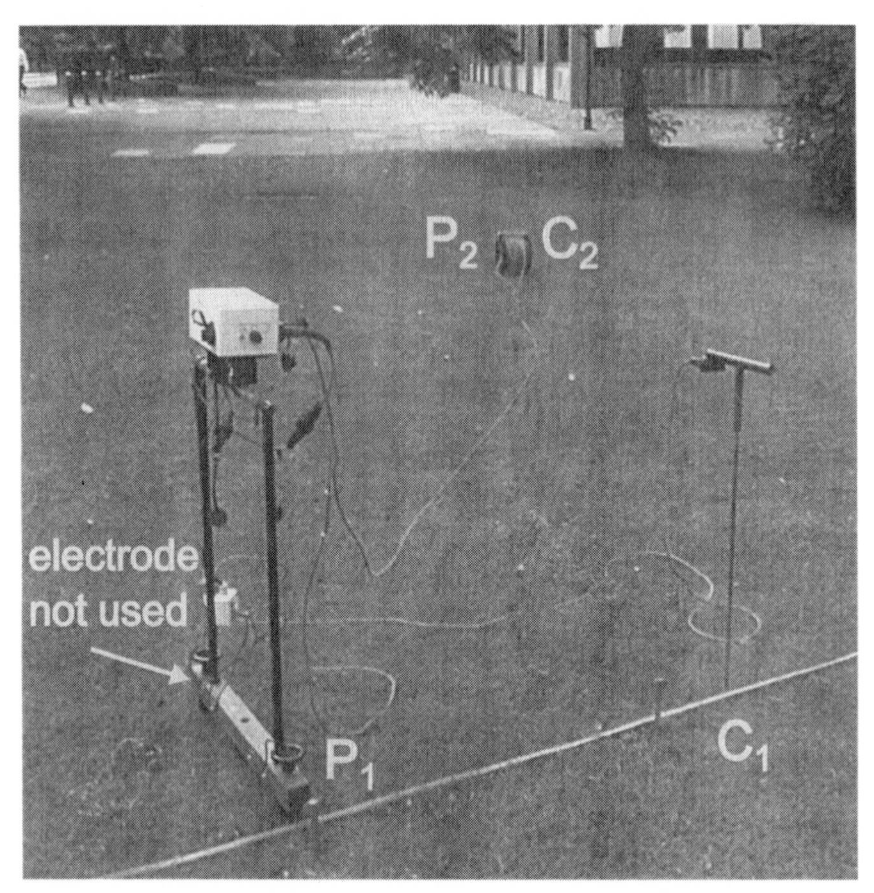

Figure 4.5. Using an RM15 it is possible to collect twin-probe ERI data with just one movable electrode that automatically triggers data recording when inserted into the ground.

introduced. These consist of arrays of many electrodes (e.g., sixty-four, or even more) that are inserted into the ground along a line at regular intervals (e.g., every 0.5 m) and are connected to a switching device that is controlled by an internal or external computer. For each measurement the switching unit selects exactly four electrodes that are used for current injection and potential measurements. The resulting basic ERI data can then be displayed as a pseudosection. There are two major designs for these systems. Either each electrode is connected to exactly one cable that is then fed to a large switching device, hence requiring a thick, multicore cable (e.g., a sixty-four-core cable), or, alternatively, "smart electrodes" are used on each metal-spike electrode, digitally controlled to connect that particular electrode to one of the four measurement terminals of the earth resistance meter. Together with two control cables, this design only requires a six-core cable between the individual electrodes. New developments use multichannel data acquisition, whereby one pair of electrodes is selected for current injection and the resulting potential on all other electrodes (e.g., sixty-two) is simultaneously measured. Then two other electrodes are chosen for the current injection and the process repeated. In this way all possible combinations of current and potential electrodes can be selected and a large number of measurements collected quickly. However, the ERI results of such a device are no longer "basic" and cannot be displayed as pseudosections. These systems collect tomographic ERI data that are discussed in section 4.6.

4.2.1 Limitations

It was outlined above that the display of data as pseudosections is based on wrongly equating the measurements' apparent resistivities with the ground's resistivities at a certain depth. Not only does this assumption discount the influence of all neighboring materials on each measurement, but it also ignores the characteristics of the electrode array used. As was shown in section 3.4.2, a gradually expanding twin-probe or Wenner array (see figures 3.13 and 3.14, respectively) will show very complex patterns of the response to the sides of a localized feature. With a twin-probe array that has its center directly over a conducting sphere ($X = 0$, figure 3.13), the resistivity response changes from a trough to a peak for a certain electrode separation. The resistivity results are shown in these figures as individual traces for different electrode separations, but they can also be displayed as a pseudosection (figure 4.6), where each trace for a certain electrode separation A is plotted as a row of grayscale data at an appropriate depth (e.g., $d = 0.5a$; $a = A \cdot r$). It can be seen clearly that the off-center resistivity variations produce "side-wings" in the pseudosection, and an inverted response is apparent "below" the actual sphere. For this well-delineated feature the theoretically calculated twin-probe pseudosection is only a very approximate representation of the actual resistivity distribution. However, for more heterogeneous resistivity distributions, even those of the test tank (see figure 4.4), a pseudosection plot of the data

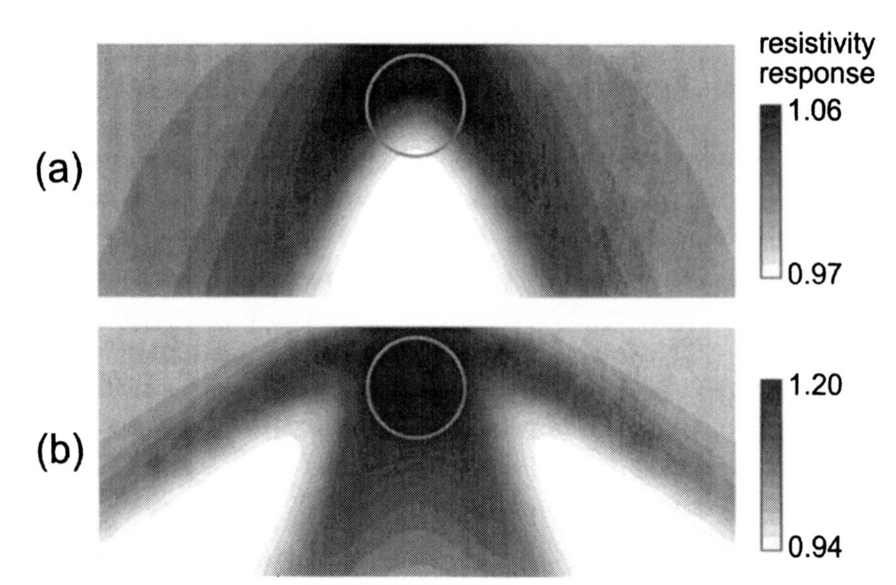

Figure 4.6. Calculated pseudosections for a longitudinal ERI over a sphere (gray circle) at relative depth Z **= 1.2. The relative electrode separation** A **increases from 0.5 to 5.0: (a) twin-probe array; (b) Wenner array.**

often creates a good visualization of buried features. When interpreting such apparent resistivity data plots, one has to be aware of the side-wings created by discrete features.

As mentioned in section 3.4, the resistivity anomalies produced when traversing a localized feature with a transverse electrode array ("broadside traverse") can be much simpler than for longitudinal traverses, and the same also applies to the display of data as pseudosections. As explained above, a transverse ERI requires "fixed center" expansions of electrodes perpendicular to the profile and is hence more time-consuming to collect. However, as can be seen from the theoretical result of figure 4.7, the shape of the buried feature is well represented in the resulting pseudosection.

Another important factor affecting ERI data is topography. A simple approach to dealing with an imaging profile that runs over undulating ground is to bend the pseudosection image according to the topographic changes, or to raise the data column at each location according to its topographic height, similar to the conventional approach applied to ground penetrating radar (GPR) data. However, if the change of topography is fairly abrupt (figure 4.8), such a change may introduce anomalies in the

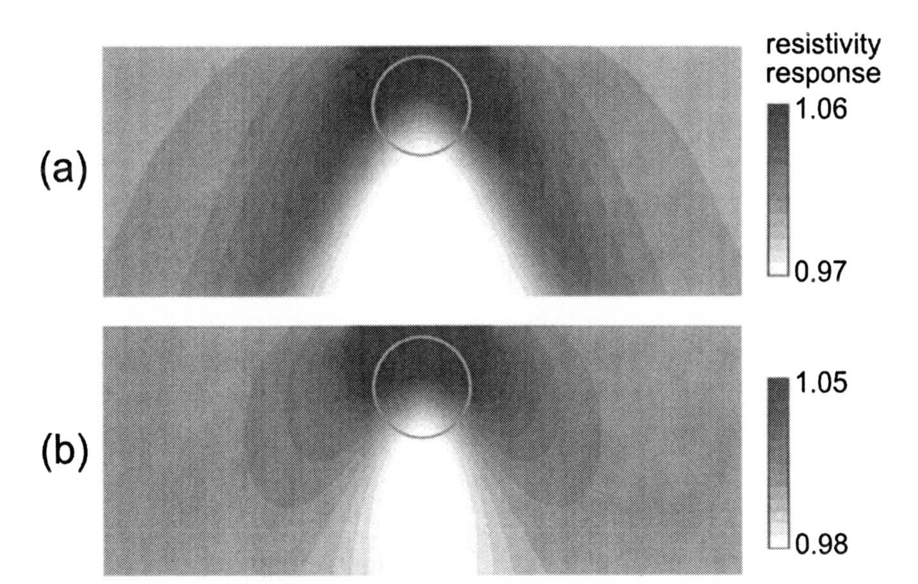

Figure 4.7. Calculated pseudosections for a twin-probe ERI over a sphere (gray circle) at relative depth Z **= 1.2. The relative electrode separation** A **increases from 0.5 to 5.0: (a) longitudinal ERI; (b) transverse ERI.**

data even if the ground underneath is entirely homogeneous. The main reason for this effect is that the current has to flow parallel to the ground surface underneath the two angled sections (see section 1.2), affecting the current density especially at the junction. In these cases complicated corrections have to be calculated (Fox et al. 1980; Yilmaz and Coskun 2011). However, as long as the topographic variations remain reasonably smooth,

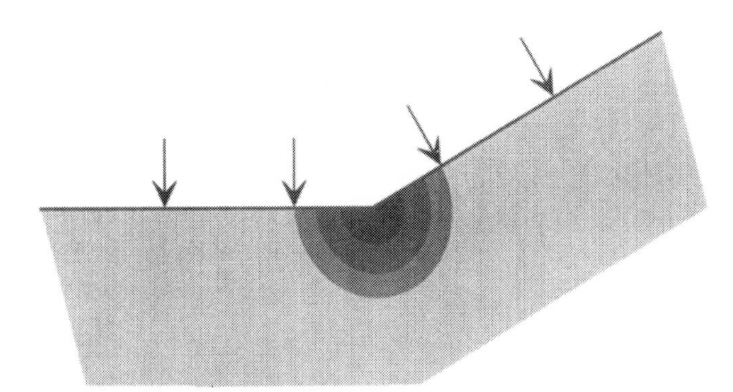

Figure 4.8. Abrupt changes of slope can create anomalies in ERI data.

as is typical over soil-covered archaeological monuments (e.g., burial mounds), this effect is shown to be negligible (Sutherland et al. 1998). At the site of Old Scatness, Shetland, for example, this information was crucial for interpreting the high apparent resistivity anomaly at the bottom of the slope correctly (figure 4.9). Although this could have been an artifact of the topographic changes, analysis of the slopes led to an interpretation of the anomalies as high resistivity stone structures (Schmidt et al. 2006). Later excavation confirmed this to be caused by the stone lining in the sides of a substantial ditch. Another reason why a simple approach to topographic corrections is usually sufficient in archaeological applications is that the electrode array often only covers a small part of the topographic variation of a feature (e.g., over a mound), not the whole range of up-slope, top, and down-slope.

4.2.2 Examples

At the graveyard site near Ripley (North Yorkshire, United Kingdom; see section 3.4.1), where shallow graves produced characteristic twin-probe responses of trough-peak-trough (section 3.4.1, figure 3.11), basic ERI profiles were collected with the twin-probe array from west to east (figure 4.10). In the western part the pseudosection clearly showed several isolated high-resistivity anomalies with pronounced side-wings tending outward (P1–P5). These anomalies were therefore most likely caused by isolated features. Of particular interest was the way the side-wings of different features intersected with each other, emanating from different anomalies and crossing each other nearly undisturbed. By contrast, the extended ditch in the eastern part (from a recent pipeline trench) showed as a single low-resistivity anomaly. Comparison with the twin-probe area survey provided further insights. The area survey was undertaken along traverses running west to east, with mobile electrodes of the array arranged north-south on the RM15 instrument frame. The characteristic three-part anomalies R1 and R2 were spread out north-south, since that was the direction of the longitudinal electrode alignments (see section 5.2.1 for further explanations). As discussed in section 3.4.1, the data of the area survey indicated that the features were approximately 0.5 m wide in the north-south direction. The ERI profiles were laid out along lines running west to east, at right angles to the area survey's electrode align-

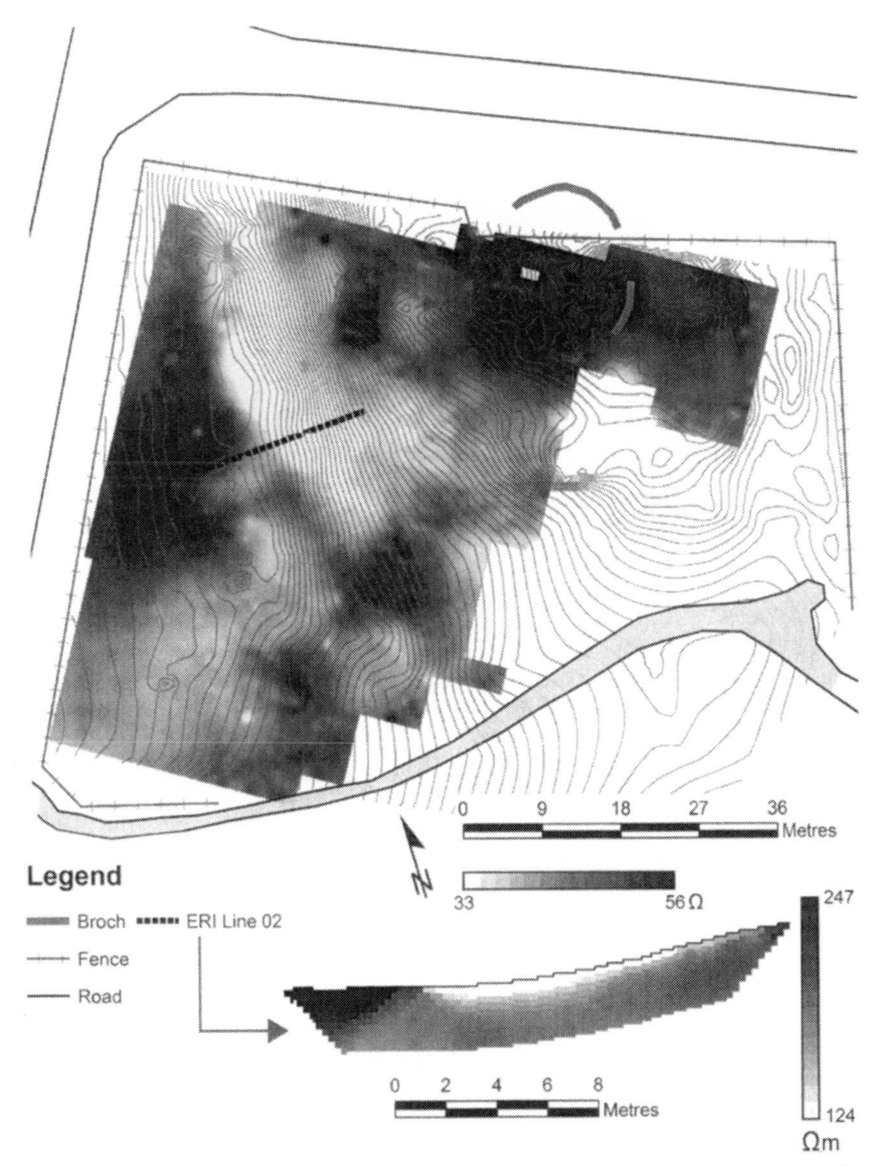

Figure 4.9. Earth resistance surveys over Scatness broch: (top) twin-probe area survey (electrode separation 1 m) with 0.1 m contours and the location of the ERI line; (bottom) twin-probe ERI investigation of the slope displayed as pseudosection.

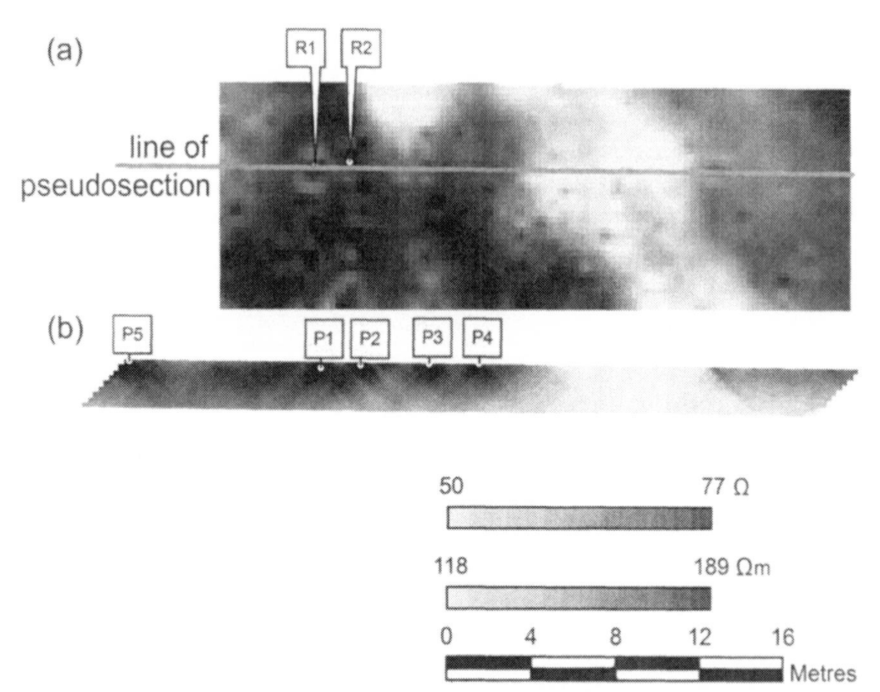

Figure 4.10. Ripley earth resistance surveys: (a) area survey with 0.5 m twin-probe array along west-to-east transects, also showing location of the ERI line; (b) twin-probe ERI displayed as pseudosection. See text for feature labels.

ments. The pseudosections clearly showed typical twin-probe side-wings ($A \approx 2$) starting from an electrode separation of 1 m. With $A = 2$ at $a = 1$ m the east-west diameter of buried localized features is 1 m, and the overall feature dimension was therefore estimated as 0.5 m × 1 m. The pseudosection anomalies P3 and P4 had only very weak or no correspondence in the area survey and appeared to be caused by deeper features. The survey demonstrated that basic ERI data can be an important supplement to area surveys, providing depth information for the interpretation of results.

In another study (Cott 1997), a ditch of the Roman Town at Caistor (Norwich, United Kingdom) was selected to monitor the influence of environmental factors on the results of twin-probe ERI profiles. Figure 4.11 shows three typical pseudosections (data from the thirteen-month cycle are discussed in section 5.3). While the location of the ditch as a low-resistance feature can be identified on all of them, its definition varies over time and was at its worst during the dry and icy January of 1996.

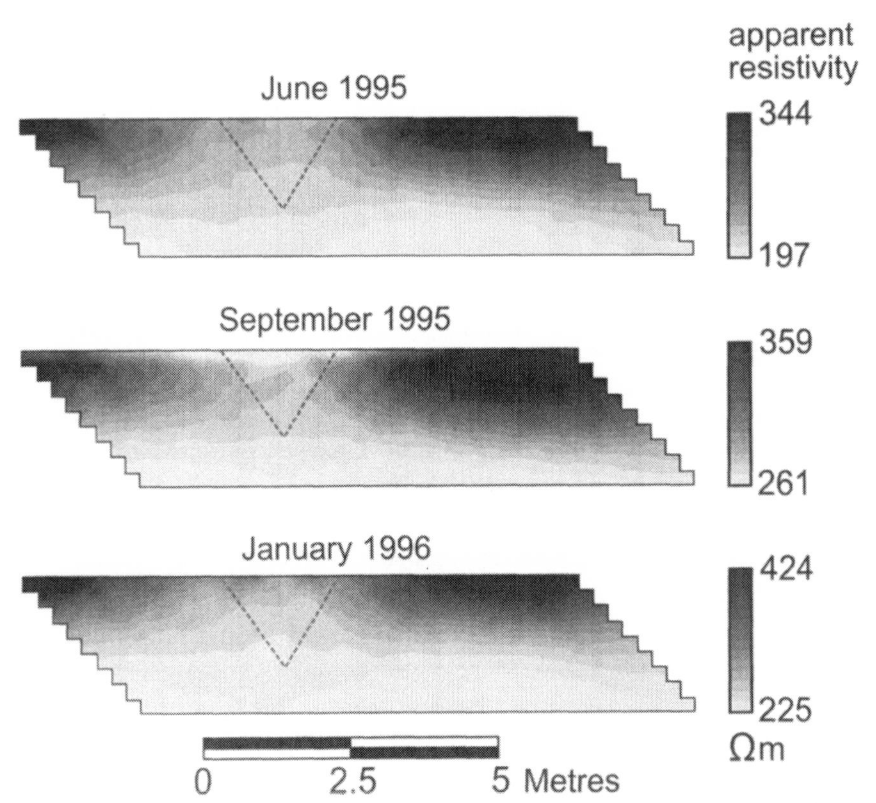

Figure 4.11. Twin-probe ERI pseudosections over a Roman ditch at Caistor, measured in different months along the same line.

While the pseudosections allow some assessment of the archaeological feature, the shape of the apparent resistivity distribution bares only vague resemblance to a typical Roman ditch, even when measured in favorable environmental conditions. This is a reflection of the limitations inherent in the construction of pseudosections.

Twin-probe ERI profiles were also measured in a modern graveyard to assess the depth of the most recent burials. This had become necessary as several grave plots were found where the top-most coffin was far shallower than the available documentation suggested, making it impossible for another burial to be interred, despite the family having paid for it years ago. Basic ERI profiles were therefore measured to obtain depth information. As shown on the pseudosection in figure 4.12, the apparent resistivity varies smoothly with burial depth, and there is no

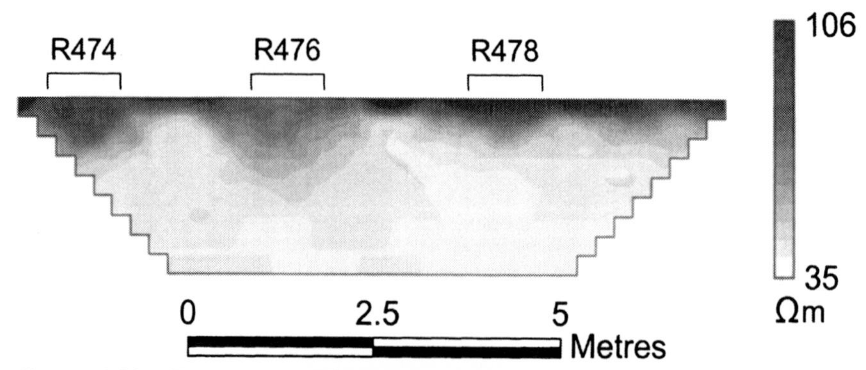

Figure 4.12. Twin-probe ERI pseudosection over three marked graves in a modern cemetery with recorded depths of 0.9 m. All three graves show clearly in the pseudosection but grave R478 appears to be shallower than stated.

sharp boundary between what appears to be the in-filled upper soil and the burial. This is partly due to the decomposition of the coffin and the resulting breakdown of the sharp resistivity contrast that such an object would have created, as well as the inherently smooth appearance of pseudosections and basic ERI data. To overcome this problem the resistivity depth profiles were calibrated using more recent graves where there was certainty about the depth of the shallowest burial. Depth estimates were then obtained from this calibration, with an error margin of approximately 0.1 m. This was a composite of errors: those due to the calibration procedure, as well as estimation errors resulting from the gradual resistivity variation in the ground. Using this method it was possible to provide estimates for the burial depths of several grave plots so that the historic records could be updated.

As part of the Yorkshire Dales Hunter-Gatherer Mobility Project (Donahue and Lovis 2006), basic ERI profiles were measured to investigate the underlying palaeolandscape. Since the remains of Mesolithic sites are usually not directly detectable with geophysical methods, a landscape approach was adopted. ERI profiles were measured in the valley of the river Skirfare (North Yorkshire, United Kingdom) to find the earlier course of the river and hence likely locations for past human settlements. This information was included in a statistical assessment of the area that also used transects of archaeological test pits. To find the best possible basic ERI configuration for this investigation, three different electrode ar-

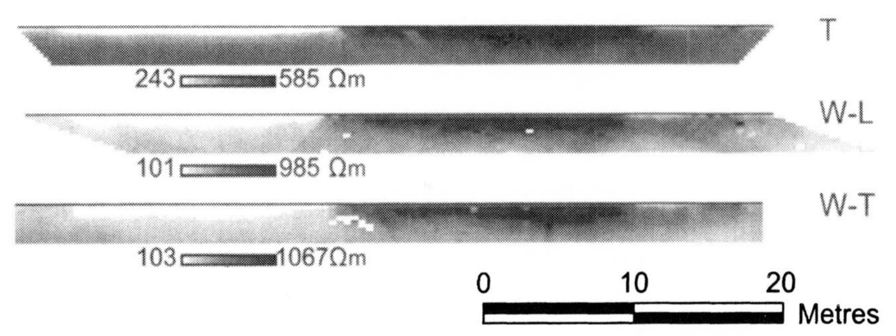

Figure 4.13. Pseudosections from one ERI transect in the Skirfare Valley with three different electrode arrays: (T) twin-probe; (W-L) Wenner longitudinal; (W-T) Wenner transverse.

rays were used along the same profile: twin-probe, Wenner longitudinal, and Wenner transverse (figure 4.13). While the last one was more work intensive (see above), it provided the clearest pseudosection results. All three pseudosection images clearly showed the conducting cross section of the old river bed on the left and a gravel lens with high resistivity on the right.

4.3 Inversion

To improve the interpretation of ERI results, mathematical methods are available that convert pseudosection images of apparent resistivities (i.e., appropriately scaled earth resistance measurement values) into actual depth sections of estimated subsurface resistivity. This process is known as inversion, which means the conversion of measured results into estimated soil property values. Various methods have been suggested (Szymanski and Tsourlos 1993), and algorithms based on Zohdy's method (see section 3.5.2) became very popular (Barker 1992; Griffiths and Barker 1994; Loke and Barker 1995a, b; 1996). In particular, the possibility of downloading Geotomo's Res2DInv software and running it in "demo mode" allowed many users to get acquainted with resistivity inversion and to evaluate its benefits. However, this software, by M. H. Loke from Geotomo Software, uses a comprehensive set of inversion parameters that are not always easy to optimize for a new user, and not all published archaeological inversion results created with this software

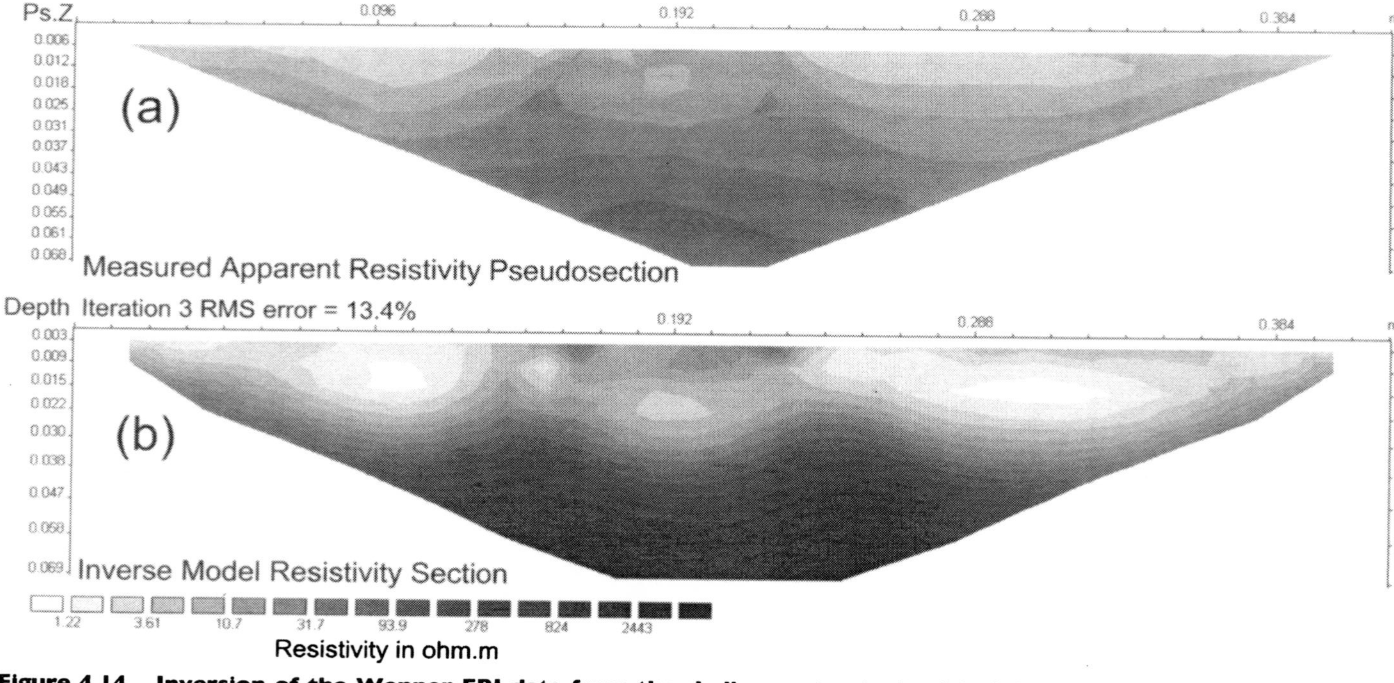

Figure 4.14. Inversion of the Wenner ERI data from the shallow water tank with default parameters in Res2DInv in the "SemiDemo" version: (a) pseudosection generated by Res2DInv; (b) inversion results of estimated ground resistivity values after three iterations.

are convincing. Figure 4.14 is the output obtained for the data from the shallow water tank (see figure 4.4 for comparison with the initial pseudosection representation of the data).

The top panel shows the software's visualization of the measured data as a pseudosection, and the bottom panel is the inversion result after three iterations with default parameters, intended to approximate the true vertical resistivity distribution. Using these default parameters the resemblance with the actual model tank is poor, and the original pseudosection display of figure 4.4 is probably a better representation. The underlying problem is the "nonuniqueness" of any inversion as discussed in section 3.5.2. A large number of subsurface resistivity distributions can be constructed that could all be used to reproduce the same measurements at the surface. The inversion therefore has to make certain assumptions in order to pick one of the various possible solutions. The first decision is the selection of a mesh of subsurface cells for which the resistivity is to be calculated. However, since the actual shape and depth of buried features is usually unknown, the default mesh of the given software implementation is normally used, even if it may not be the most appropriate choice.[2] Even more important is the selection of parameters that govern the smoothness of inversion results. While in many geological and environmental applications a relatively smooth variation of resistivity values in the ground is likely, for archaeological applications fairly abrupt changes and vertical edges are to be expected—for example, when investigating buried walls or ditches. The smoothness constraints used in the Res2DInv software are not necessarily the best choice in such circumstances. Furthermore, Loke et al. (2003) found that the software's use of the L2-norm for the inversion leads to overly smooth variations in the resistivity values and that the L1-norm may be better suited in such instances, creating a robust or blocky inversion.[3] The issue is further exacerbated by the frequent use of the software's standard color scheme for plotting its results.[4] This color scheme

2. After a first approximate inversion a refined mesh could be selected by the user to improve the results.

3. L1-norm and L2-norm refers to the way in which the overall discrepancy is calculated between the measured apparent resistivity values and the resistivity values that a given subsurface model would produce. The L1-norm is the sum of all absolute values of differences, whereas the L2-norm is the square root of the sum of the squared differences.

4. The full version allows exporting the inversion results for display in other visualization packages.

introduces fairly abrupt transitions between adjacent resistivity values, and selecting a grayscale range as in figure 4.14 may produce better results. Alternative inversion methods were investigated by other researchers and often produce very good results for archaeological features. For example, Tsourlos et al. (2005) were able to calculate inversion results with a generalized back-projection algorithm for the guest hall at Fountains Abbey (Yorkshire, United Kingdom) that very closely resembled the foundation walls, known to be still in the ground. There also exist algorithms for the three-dimensional inversion of data, which will be discussed in the next section, and tomography-based schemes that are highlighted in section 4.6.

It is clear from this discussion that a large number of parameters exist that can be adjusted and different algorithms that can be used to calculate a resistivity inversion. Although a statistical evaluation of all the possible results would be possible (e.g., using Monte-Carlo simulations), it is often the case that simply the "best looking" inversion is selected as a final result. Such a selection is a subjective choice and involves the interpretation of measurement data, usually by taking archaeological background information about the expected outcomes into account and then selecting an inversion that shows these expected outcomes best. Given the non-uniqueness of resistivity inversions, results should therefore not be regarded as a "scientific fact" but as one of many possible solutions, based on a deliberate, or possibly unreflected, choice of parameters and algorithms. By contrast, pseudosections are just a display of measurement values with minimal adjustment of parameters (e.g., depth scales) and are hence far less affected by subjective parameter choices. This situation is analogous to the difference between a display of raw data from an earth resistance area survey and its representation as a heavily filtered data plot that was created to highlight certain aspects of the data chosen by the interpreter (see chapter 6).

Although resistivity inversion does not always improve the definition of the shape of buried features, as shown above, it often allows a better estimate of the actual resistivity values. This is of particular importance in environmental applications, where the exact resistivity is often needed to assess levels of contamination or soil moisture. In archaeological applications, however, soil resistivity usually varies considerably due to changes in moisture levels, and absolute values are far less important. Instead, it is the shape of boundaries between features with different resistivities that is the most important information to be deduced from ERI data.

4.4 Pseudoslices

Basic ERI profiles and their display as pseudosections provide a view into the ground as vertical sections. For features that are defined by a distinct cross section this is a valuable insight. However, where features are better characterized through their lateral extent a different approach may be more suitable. Instead of acquiring just one single, basic ERI profile across a feature, measurements along many parallel lines can be recorded. The same data that would then usually be displayed in a series of vertical sections can also be used to form horizontal slices of apparent resistivity at different pseudodepths, so-called pseudoslices (figure 4.15).

A survey undertaken at the Roman fort of Newstead (Melrose, United Kingdom) by Paul Cheetham and Rick Jones showed the respective strengths of pseudosections and pseudoslices (figure 4.16). The sections very clearly delineated the cross section of a pit (which was confirmed through later excavation). By contrast, the slices were better suited to show the course of a Roman road, to the right, and for the delineation of linear foundations, visible as two thin, parallel, horizontal resistance anomalies, in the middle. The complementary nature of the two views helped to assess the information contained in the data.

Earth resistance surveys are usually undertaken to map subsurface features over an area. The electrode array for such an area survey is then mounted on a frame and moved to different locations to record the earth resistance, which is subsequently mapped in a two-dimensional horizontal plot. Usually one electrode configuration is used for the whole area,

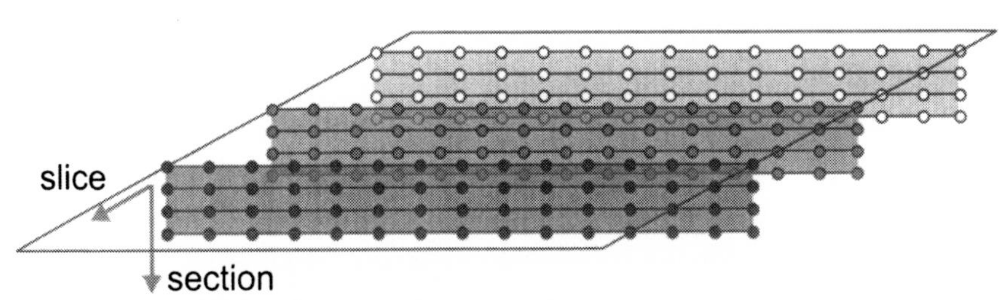

Figure 4.15. Data from several parallel ERI profiles can be displayed either as pseudosections or as pseudoslices.

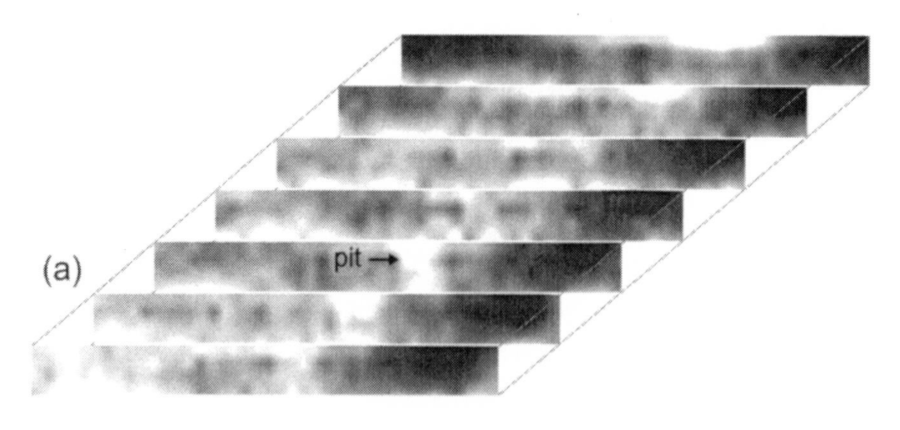

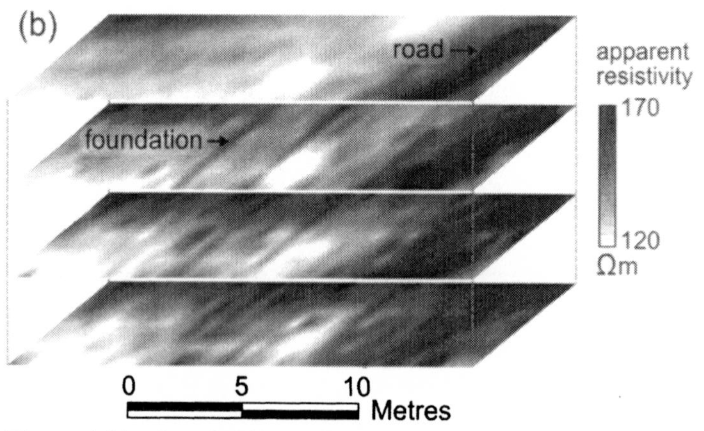

Figure 4.16. Longitudinal "fixed center" twin-probe ERI data collected at the Roman fort in Newstead, showing a pit, a road, and two lines of parallel foundations: (a) the data displayed as seven parallel pseudosections; (b) the same data visualized as four horizontal pseudoslices.

leading to a single earth resistance map. However, the combination of an RM15 earth resistance meter and an MPX15 multiplexer from Geoscan Research (or the combined RM85) allows using up to six electrodes on a frame that can be switched into different arrangements. It is then possible to undertake twin-probe measurements and record readings for six different separations of the mobile electrodes (figure 4.17). When such readings are collected along a line, the results from the different electrode separations can be assembled into an ERI and displayed as pseudosections. Undertaking a whole area survey with this device allows sorting the data into pseudoslices that can be plotted for each individual electrode

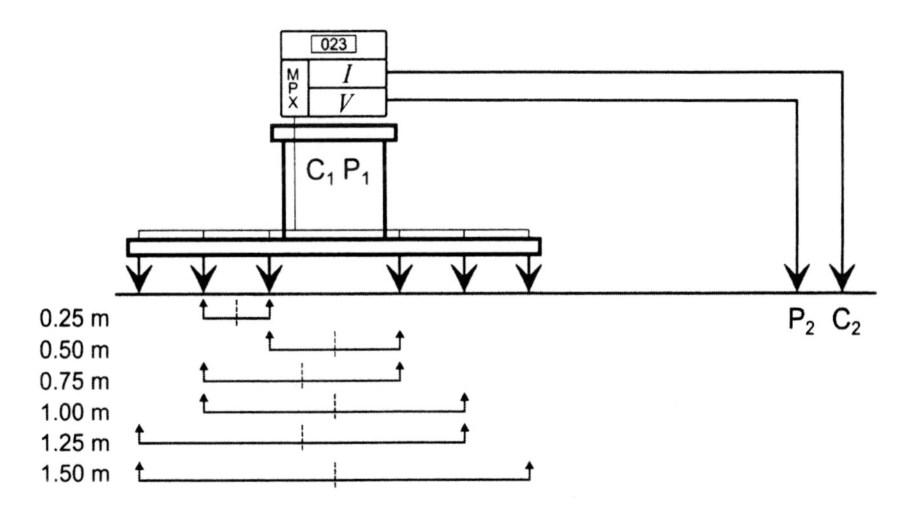

Figure 4.17. Using a multiplexer (MPX) connected to six electrodes it is possible to make measurements with six different twin-probe separations in a sequence. The center of the first measurement with 0.25 m electrode separation is 0.375 m from the center of the electrode frame, which has to be corrected by appropriate processing. The maximum lateral shift of the other five measurements is only 0.125 m and can be ignored.

separation, effectively depicting features at different depth. In a survey at the Roman town of Wroxeter (United Kingdom), this approach helped to investigate the depth range of different components of a Roman building (figure 4.18).

There is another possible use for measurements collected with different twin-probe separations in the same location. Often the separation between the two remote electrodes is not measured, which makes it nearly impossible to calculate the correct apparent resistivity of the results according to equation 2.16, as would be required for a comparison of surveys from different sites (see section 5.2). Using equation 2.16 and two different separations of the mobile electrodes (a_1 and a_2), the two resulting resistance measurements (R_1 and R_2) allow the calculation of the correct apparent resistivity independent of the separation between the remote electrodes.

$$\rho_A = 2\pi(R_1 - R_2)\frac{a_1 a_2}{a_2 - a_1} \tag{4.2}$$

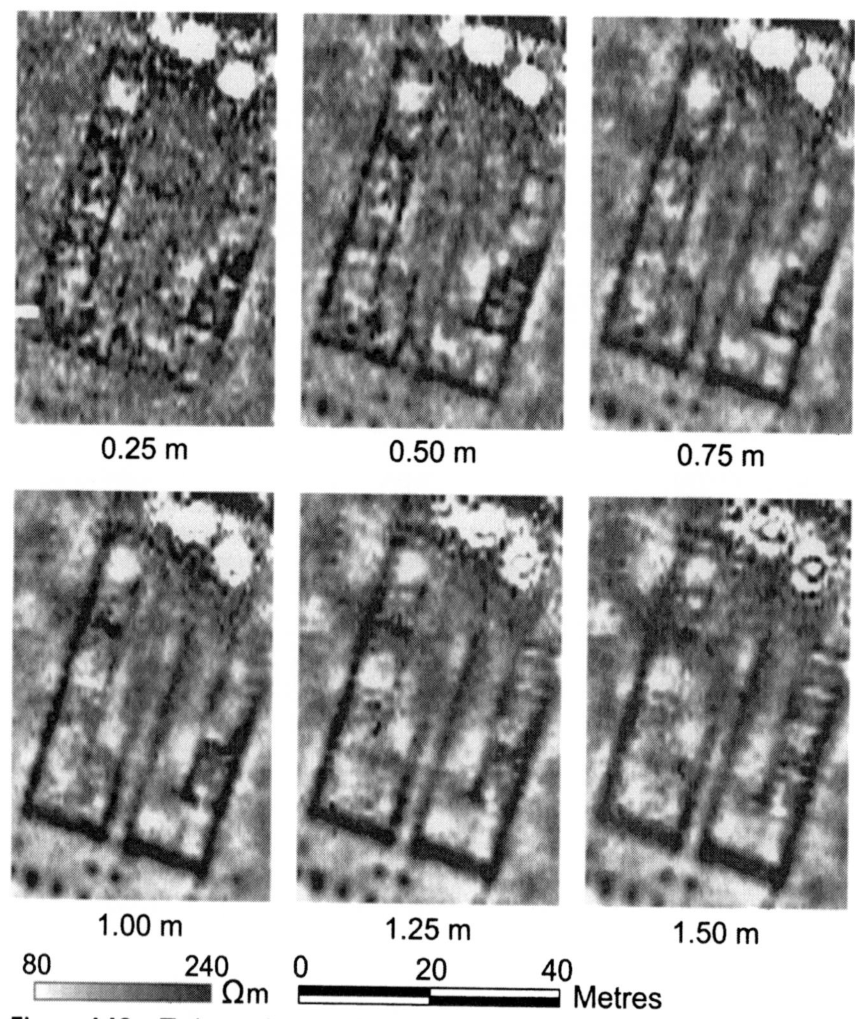

Figure 4.18. Twin-probe pseudoslices over the remains of a Roman building at Wroxeter. The data were collected with an RM15 and an MPX15 to measure six different twin-probe electrode separations simultaneously. Measurements were converted to apparent resistivity and high-pass filtered to show the outline of the building more clearly. The data indicate that some internal walls and the column bases at the front do not extend over the full depth range. Data courtesy of Geoscan Research.

If the two mobile electrodes' separations are 0.5 m and 1.0 m, respectively, Clark (1996: 46) noted that the equation simplifies to

$$\rho_A = 2\pi(R_1 - R_2) \cdot 1\text{m}, \tag{4.3}$$

where 1 m is included only to yield the formally correct units of the result.[5] Thus subtracting the two datasets from each other and multiplying with 6.2 produces an apparent resistivity map that can be compared even if the remote electrodes had different separations on different sites. However, this calculation is only correct for the homogeneous halfspace, and in most cases the two measurements will probe resistivities at different depths of the heterogeneous ground.

Pseudoslices can be a very useful tool for environmental investigations. For instance, to monitor the confinement of pollutants in landfill sites pseudosections can be recorded at regular time intervals. If a plume of pollutants develops from the site, its ionic content will reduce the electrical resistivity of the ground considerably. Pseudoslices created from such data are therefore a useful tool for environmental monitoring. For example, the pseudoslices depicted by Burger and Burger (1992) clearly showed the extent of the contaminating plume. In a similar way, pseudoslices can be used to map the lateral extent of aquifers, and good results of this method were, for example, obtained in Sweden (Christensen and Sørensen 1998).

As stated above, pseudoslices are created from the same data that are usually displayed as vertical sections. It is therefore just another step to use all the data and display them in three dimensions. Modern software allows the calculation of three-dimensional iso-surfaces of constant apparent resistivity that can be displayed as three-dimensional images. Figure 4.19 shows the data from the Roman fort at Newstead (see figure 4.16) with different iso-surfaces. At a level of 164 Ωm the Roman road on the right edge is clearly visible. However, the linear anomalies of the foundation in the middle of the survey area are not well represented (145 and 148 Ωm), as the gridding algorithm for the visualization created isolated anomalies between the vertical sections that are 0.5 m apart. The pit in the front of the investigated block shows different aspects of

5. R is in Ω, while ρ has to be in Ωm.

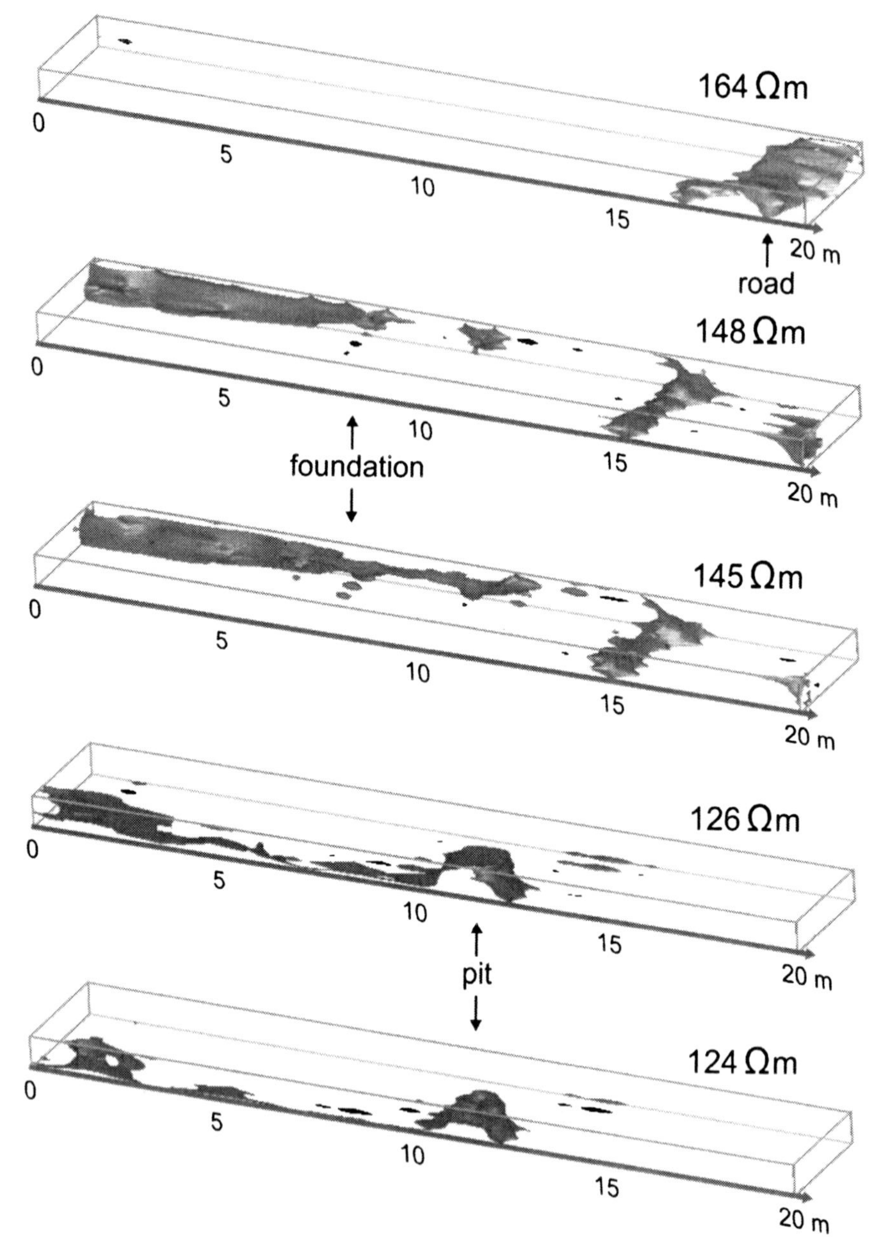

Figure 4.19. Three-dimensional visualization of the twin-probe ERI data from Newstead Roman fort. To show the features identified in the pseudosections and pseudoslices of figure 4.16, several apparent resistivity isosurfaces have to be examined.

its shape at 124 and 126 Ωm thresholds. Although these resulting three-dimensional images are pretty, they are not easily interpreted, as different threshold values have to be tested to explore various aspects of the data. The main reason for this is the smooth variation of apparent resistivity values, which means that a single iso-surface is not necessarily representative for a feature over its full volume extent. Grayscale or color images of two-dimensional sections and slices may allow a more comprehensive interpretation and should also be consulted.

4.5 Three-dimensional inversion

Data that were collected along several parallel profiles as individual basic ERI sections cannot only be displayed as pseudoslices, but can also be inverted with three-dimensional algorithms. An implicit assumption for a two-dimensional inversion of ERI data is that the soil resistivities in the direction perpendicular to the ERI profile do not vary, having the same depth distribution as over the measured profile. This simplistic view of soil properties in three dimensions is often referred to as 2.5D (i.e., "not quite 3D"), and it describes only a few features correctly—for example, a long ditch with homogeneous fill in homogeneous surrounding soil. When data for several parallel profiles exist they can be jointly inverted, creating a full three-dimensional dataset of subsoil resistivities. Papadopoulos et al. (2006) tested several different approaches for dealing with such datasets. The simplest method is to invert all ERI sections individually in two dimensions and then build a three-dimensional volume of resistivity data by appropriately assembling the individual results in three dimensions. A refinement is to extract additional artificial ERI profiles from the data that are perpendicular to the actual measured sections (they are usually referred to as "along the y-axis," where the x-axis is the original profile direction). These extracted y-direction data are then also inverted in two dimensions, and the results are combined with the inversion results from the original x-direction profiles—for example, by averaging both in three dimensions. The most advanced method is to use a full three-dimensional inversion algorithm to obtain the ground resistivities from all individual measurements. Papadopoulos et al. (2006) found that combining the two-dimensional inversion results from the two perpendicular directions created resistivity data that represented the assumed archaeological features very well, and certainly better than when inverting only individual profiles. The full three-dimensional inversion improved the results even further, albeit only incrementally. Based on these findings it seems well worth the extra effort to collect data in such a way that advanced inversion schemes for three-dimensional data can be used. Expanding on this idea, it is possible to use a two-dimensional grid of surface electrodes over the whole feature area to inject current and to measure potential between all possible combinations of electrodes. Such data recording is referred to as three-dimensional tomographic ERI measurement.

4.6 Resistivity tomography

To overcome some of the problems associated with the interpretation and inversion of basic ERI data, resistivity tomography[6] has been developed. Tomography is best known from the medical realm, where CT scans (i.e., Computer Tomography scans) use X-rays from different directions for the computer reconstruction of a human body's interior organs.

In a conventional X-radiograph a broad sweep of parallel X-rays is used to illuminate body parts from the front and record the transmitted rays at the back, either on a radiographic film or with electronic detectors. This record shows the cumulative absorption that the X-rays encounter along their path through the body. It is not possible to derive depth information from these conventional X-radiographs. In X-ray tomography the transmitters and receivers are rotated around the investigated body to produce absorption data for different pathways (figure 4.20). After combining the results from these different pathways it is possible to resolve the spatial distribution of X-ray absorption for the whole cross section and thereby record the location of different body parts. For example, feature (A) in figure 4.20 is hit by ray 2 in recording position (a) and by rays 2 and 3 in recording position (b). Through a computational process—for example, iterative back-propagation—it is possible to calculate the increased absorption at location (A). In such a way, two-dimensional slices through the investigated feature can be constructed. In a CT scan this is then

6. Also referred to as ERI tomography, tomographic ERI, or Electrical Resistivity Tomography (ERT).

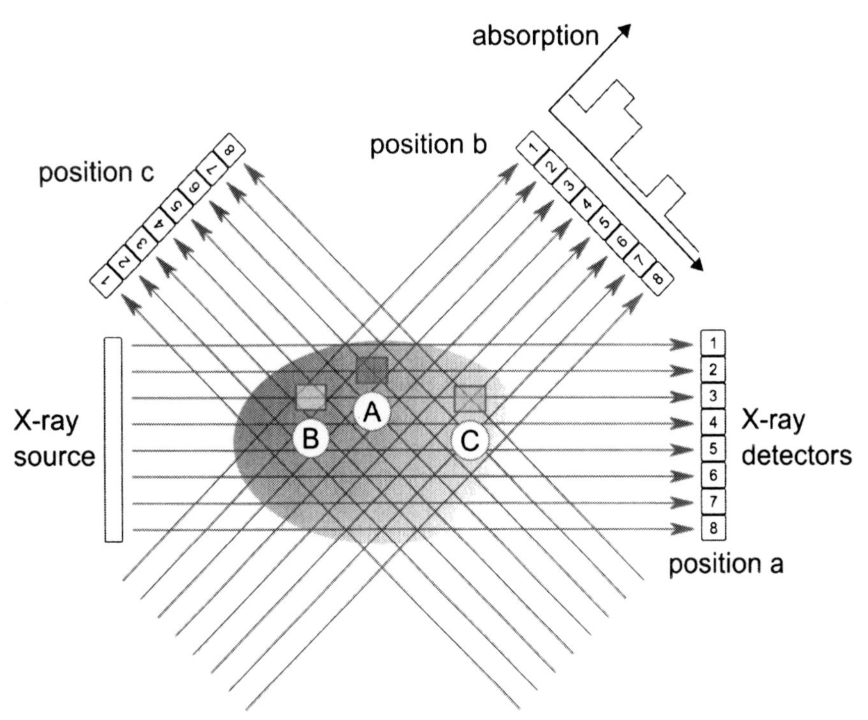

Figure 4.20. For X-ray tomography the position of the X-ray source and detectors is changed (positions a–c) to illuminate internal body parts (A–C) from different directions. The two-dimensional distribution of absorption can be computed from the combined results of all directions.

repeated for many closely spaced slices to generate a three-dimensional volume of data.

Tomography can also be applied to earth resistance measurements, leading to the technique of "resistivity tomography." The paths of X-rays are replaced by the curvy current lines, and the different orientations of X-ray source and detector (e.g., (a) and (b) in figure 4.20) are implemented through different positions of the electrodes. To achieve this, a considerable number of electrodes have to be placed in the ground and automatically switched to measure earth resistance in all possible configurations (not just as a single type of array—for example, Wenner). The resulting data are then processed, and a true resistivity section is calculated (Noel and Xu 1991; Szymanski and Tsourlos 1993). In contrast to X-ray tomography, current lines are not straight, and when passing through features of interest they are not just weakened but also deflected. Consequently, the mathematical treatment is much more complicated, and new algorithms had to be found.

It was mentioned above that even basic ERI profiles are sometimes recorded by connecting a number of electrodes to an earth resistance meter via appropriate cables and switches. The difference to a tomographic measurement is that in a basic ERI only one type of electrode array (e.g., Wenner or twin-probe) is used along the whole profile. This makes it possible to display the data as a pseudosection without further processing. However, no such simple display can be found for the large volume of data that is collected in a tomographic ERI. Kampke (1999) showed how basic ERI profiles from different arrays can be combined to create a better definition of underlying features, thus developing a hybrid between pseudosections and resistivity tomography using several electrode configurations.

With modern switching technologies it is possible to measure apparent resistivities between hundreds of electrodes, placing them over and around features of interest as a grid, in a circle or in any other configuration. The large volume of data collected with such systems requires sophisticated processing methods and has only become feasible with recent advances in computer power.

CHAPTER FIVE
IMPLICATIONS FOR FIELD PRACTICE

In previous chapters various topics on earth resistance measurements were discussed and illustrated with examples. The insights gained will be used in this chapter to examine issues relevant for field practice, mainly in the context of archaeological geophysical prospecting.

5.1 Measuring earth resistance

To measure earth resistance, a current I is passed through the ground, and the resulting potential difference ΔV is registered. From these two quantities the earth resistance R is calculated (see equation 1.2). Often the current source and the device that measures the potential difference are incorporated into the same instrument ("earth resistance meter"), but some devices that were originally designed for geological applications use large external batteries. In the following sections, the aspects related to current source and potential measurement will be discussed separately.

5.1.1 Current sources

Considering typical resistivities for wet and dry soil, 0.60 Ωm and 16.7 Ωm, respectively (see table 1.2), measurements with an ideal twin-probe array of 0.5 m electrode separation yield an earth resistance of 0.38 Ω and 10.6 Ω, respectively (see equation 2.17).[1] A limestone foundation

1. It is interesting that measurements on a "geological scale" with an electrode separation of $a \approx 10$ m would yield resistances as low as 0.02 Ω and 0.53 Ω.

wall may be taken as an example for a high-resistance anomaly. Resistivities between 400 Ωm and 700 Ωm (wet and dry, respectively; see table 1.2) produce corresponding earth resistance measurements between 255 Ω and 446 Ω. The typical range of readings to be expected in a survey will hence lie somewhere between 1 Ω and 500 Ω. Considering that modern electronic circuitry can easily measure voltages between 1 mV and 500 mV it is clear that a current of about 1 mA is already well suited for archaeological earth resistance measurements (see equation 1.2). Given these parameters the power required from the current source is below[2] 0.5 mW, and it is therefore possible to use standard dry cells or chargeable batteries as supplies for the earth current. Such battery-powered devices are easily portable and can last for more than a survey day. In some non-archaeological applications (e.g., for deep geological features) where far smaller resistances have to be measured, higher output power is required and more substantial power supplies are needed. For example, if a 0.1 Ω earth resistance were to be measured with 500 mV, the current would have to be 5 A and the power supply would require an output of 2.5 W.

The discussion in section 1.3.2 highlighted the requirement that a current source has to switch its polarity frequently to avoid the accumulation of charged ions around the electrodes and hence their screening. Between the polarity reversals the current is kept constant to avoid electromagnetic effects (i.e., induction) or polarization. Such a current is referred to as switched direct current (switched DC, section 1.3.2). The repetition rate of the polarity switching can sometimes be set by the user to an "unusual" frequency (e.g., 137 Hz) to distinguish the measurements from effects created by other current sources (e.g., power lines at 50 or 60 Hz). If the circuitry for the measurement of the potential difference is tuned to exactly the same frequency as the switched current, interferences from unwanted signal sources can be minimized.

In addition to these currents sent through the electrodes C_1 and C_2 by the power supply, there are also natural currents in the soil that can be exploited for earth resistance measurements. Telluric currents are induced by variations of the magnetic field around the earth. Such variations are mainly caused by the solar wind leading to currents at great depth that could not have been injected by a power supply and provide a

2. 500 mV × 1 mA = 0.5 mW

convenient source for deep investigations. Due to their very nature they cannot be controlled and may only be available intermittently. Their use for archaeological prospecting is therefore limited. More reliable sources are ore bodies with different electrochemical potentials that lie close to each other. They can act as "batteries" for natural currents that lead to a so-called self-potential at the surface. There was some initial success for archaeological investigations using this method in the 1980s (Wynn and Sherwood 1984; Wynn 1986), but it is now rarely used due to the difficulties of interpreting data and the necessity to use nonpolarizable electrodes, which makes data collection cumbersome at high spatial resolution. Nevertheless, some interesting results from this technique are still being reported (Drahor et al. 1996; Drahor 2004).

5.1.2 Measurements of electrical potential

For the measurement of electrical potential differences many instruments have been developed, which are often referred to as voltmeters. This section considers which specific requirements are needed for earth resistance measurements. It was mentioned above that a suppression of unwanted signals ("noise") can be achieved by linking the measurement of the potential difference to the switching frequency of the applied DC current. However, even more important are efforts to reduce the effect of "contact resistance."

So far it has tacitly been assumed that four electrodes are required for the measurement of earth resistance. However, this needs further justification since it might be possible to construct a circuit with just two electrodes (figure 5.1a) that could potentially be used for these measurements. The problem encountered with such an arrangement is the effect of the contact resistance R_c. This is due to the usually limited contact between a metal electrode inserted into the ground and the surrounding soil. The current has to pass through usually considerable gaps and pores, which leads to high and possibly variable values of R_c in the order of about 1 kΩ to 1 MΩ. In an arrangement as depicted in figure 5.1a, the measurement would wrongly result in a resistance of $R_c + R + R_c$ (where R is the actual earth resistance of interest), since the current required for the measurement of the potential difference is also affected by the contact resistance. Since the contact resistance is much larger than the earth resistance of interest, such a measurement would be useless.

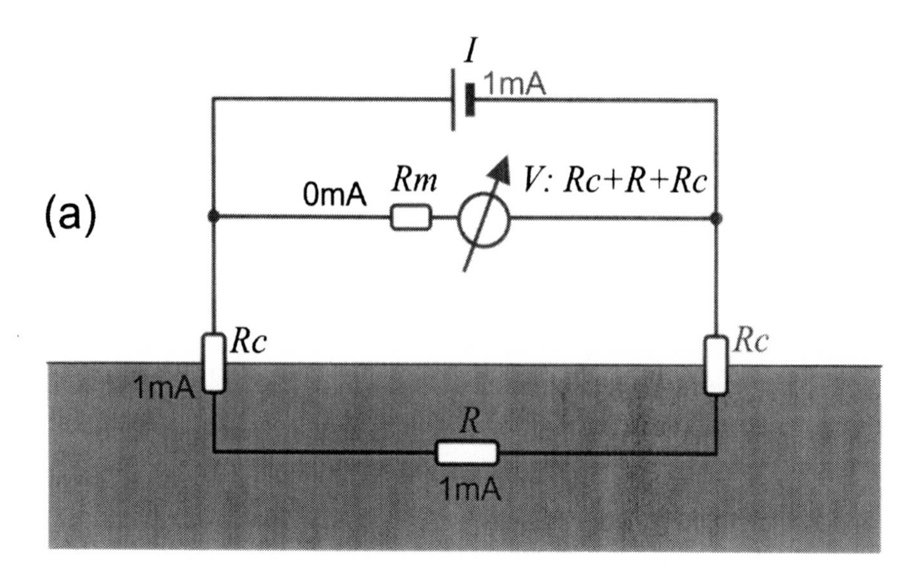

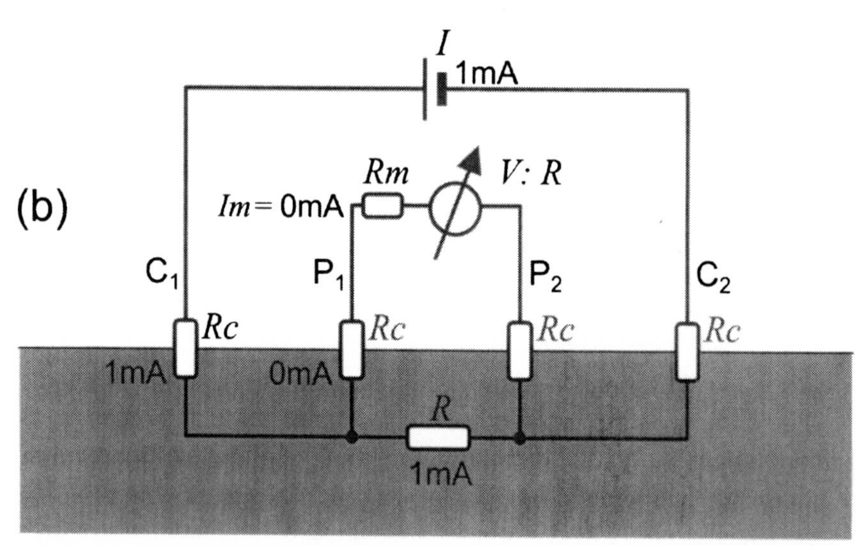

Figure 5.1. Contact resistances are encountered in earth resistance measurements: (a) if only two electrodes would be used, the voltmeter would measure the earth resistance R together with the contact resistance R_c; (b) with a four-electrode arrangement and an earth resistance meter with very high impedance R_m the influence of the contact resistance can be minimized.

To avoid this problem, four electrodes are used (figure 5.1b). In this arrangement a distinction can be made between the earth current I that produces the potential difference and the current I_m that is required to *measure* this potential difference. On its way through the ground the earth current encounters the contact resistances at the current electrodes, but this can be counteracted by the current source, which regulates the current to a constant value, irrespective of any electrical resistance it encounters. The current entering the ground is hence well defined. In order to measure the true earth resistance, it has to be ensured that all of this earth current flows through the ground and next to nothing along the measurement path that contains the contact resistance of the two potential electrodes and the voltmeter itself. This unwanted pathway forms the measuring current I_m. To facilitate this, the internal resistance of the voltmeter R_m (its so-called impedance) has to be very high, much higher than any possible earth or contact resistance. Such a high impedance will virtually suppress the measuring current I_m, and the current through the soil to be investigated will be nearly identical to the known earth current I so that the ratio between the measured voltage difference and the known earth current is indeed the sought-after earth resistance. Building a device that can measure potential differences when only a very small current is flowing through it is demanding and accounts for the much higher price of earth resistance meters compared to a simple handheld resistance meter. It is hence the combination of a four-electrode array with a high-impedance voltmeter that allows the measurement of earth resistance without having to worry too much about possible problems with contact resistances. The Geoscan RM15 instrument, for example, can usually tolerate a contact resistance of 0.1 MΩ. In some conditions, however, even this may not be sufficient, and problems were reported when surveying in the sands of Egypt (El-Gamili et al. 1999; Mathieson et al. 1999). One common solution for resistivity imaging measurements in such locations is to place a set of electrodes permanently in the sand, water them well, and cover the wet sand to stop it from drying out too quickly. However, even after watering, the contact resistance may still be too high, and the eventual evaporation of the moisture can lead to changing earth resistance readings, making earth resistance measurements in sand challenging. If electrodes are to be placed on hard surfaces—for example, the walls of old buildings or the asphalt of a car park—fairly good contact can be made using conductive bentonite clay that is applied between flat metal plates as electrodes and the surface of investigation. Although still high, the resulting contact resistance will be constant and not variable, making the interpretation and further processing of results reliable (Tsourlos and Tsokas 2011).

5.1.3 Instrument design

As there are many competing requirements, there is no "best" design of an instrument for archaeological prospecting, but a few criteria are worth considering. The following discussion draws on important aspects of earth resistance measurements that were introduced in earlier sections.

Range

Most digital instruments have the ability to adapt their input range to the prevailing earth resistance values on a site by applying an amplification to the potential difference that lies across the two voltage terminals before the actual value is measured. The amplification is often expressed as a "gain" (e.g., × 1, × 10, × 0.1, etc.). The advantage of such a gain is an improved match between the supplied voltage and the electronics of the instrument. For example, if the readings over a site vary only between 10 Ω and 20 Ω, it would be appropriate to apply a gain of × 10 to a typical instrument like the Geoscan RM15. This gain selection would allow the instrument to switch to a more sensitive measurement mode (as it does not have to cope with readings above 20 Ω) and to use the internal analog-to-digital conversion to provide additional precision by measuring an additional decimal place so that values of 12.8 Ω and 13.2 Ω could be recorded. In this example, if the gain had only been set to × 1, both readings would be recorded as 13 Ω and would therefore be undistinguishable. By contrast, on a site where the earth resistance reaches up to 300 Ω an instrument with a gain setting of × 10 would be overloaded with such a high reading, resulting in an overrange message. It is therefore important to assess the possible range of earth resistance values of a particular site in advance of a survey (e.g., by taking a couple of readings across the area) to determine the best-suited measurement gain, as some instruments do not allow changing the gain once the recording of measurements has commenced.

Current

Another parameter that can often be adjusted in earth resistance meters is the DC current used for the measurements. As outlined above, a current of 1 mA is normally appropriate. However, if very high earth and contact resistances are encountered, unacceptably high potential differences may be produced across the current source, since, according to equation 1.2, $V = (R_c + R + R_c)\, I$ (figure 5.1b). In this case it may be necessary to reduce the current to comply with the instrument's allowed rating across the current terminals (e.g., 40 V).

However, not always is a lower current advantageous in such high-resistance situations. As was shown before, the high-input impedance of an earth resistance meter normally allows neglecting effects caused by the electrodes' contact resistance. However, if the latter becomes considerably higher than the impedance, slight variations of the contact resistance (e.g., at adjacent measurement positions) will cause different readings of the meter. How, then, can the contact resistance be reduced to an acceptable level? As outlined above, one possible solution is to moisten the soil around the electrodes, but this is time-consuming, and the gradual drying out of the soil can introduce undesirable time variations of the measurements. Another option is to increase the measurement current. Due to the complex nature of the currents flowing between the soil and the electrodes (electrons from the electrode are absorbed by the ions gathering there), the contact resistance does not strictly follow Ohm's law (see section 1.1.1), but depends on the current used. A higher current usually lowers the contact resistance. Hence, to overcome contact problems higher currents may be used—but not as high as to cause excessive potential differences that could damage the instrument (see above).

Balancing these two requirements may not always be easy. Some instruments are designed to operate in a "constant voltage" mode. The simpler design of electronic circuits that maintain a constant current, regardless of external earth resistance, means that "constant current" sources are more commonly found in earth resistance meters. However, if instead the voltage across the current source is kept constant at a high but optimal level, the current can be regulated to its highest possible strength and contact resistance is minimized. Earth resistance is again determined from the measurements of voltage difference and current according to equation 1.2. Such a design is particularly useful for surveys in dry conditions.

Automatic recording

To make the operation of an earth resistance meter more convenient, automatic recording of the measurements can be implemented. When moving an electrode array from one position to the next, the current path through the ground is first interrupted and then reestablished. An instrument that detects this change of current can automatically trigger an internal microcontroller to record a new measurement. To avoid erroneous readings some further considerations are necessary. When inserting an electrode in the ground a current path may already have been created when the electrode's tip touches the soil, or when it touches wet grass. The resulting immediate measurement would not be representative of the sought-after earth resistance that can only be measured once the electrodes are fully inserted. A simple solution to this problem is a slight delay between the first electrical contact and the recording of the measurement. By the time the resistance value is logged, good contact with the ground is usually established. The delay is commonly set through the instrument's "logging speed," and therefore this has to be adjusted according to the site's condition (not too fast in the presence of wet grass, or for dry and hard soil that is difficult to penetrate) and balanced with the desire to obtain as fast a recording speed as possible.

5.1.4 Choice of array

In section 2.4 the responses of various electrode arrays were investigated in detail. The question arises as to which of them is best suited for archaeological prospecting. To make such a judgment some assessment criteria have to be defined first. The most relevant are summarized in the following list.

- ***Strength of response***: In order to detect anomalies caused by features with weak resistivity contrast, the anomaly should be as strong as possible (i.e., very high compared to the background resistance for positive/insulating anomalies and low below the background resistance for negative/conducting anomalies). Only then can the anomalies be discerned easily from the background and its variations.

- ***Spatial resolution***: Archaeological features are often found in close vicinity to each other (e.g., postholes). In such cases it is important that the measured anomalies do not overlap so that individual features can be clearly separated. The spatial extents of anomalies should therefore not be much larger than the features that are causing them.
- ***Depth distinction***: This describes how well features can be distinguished when they are buried at different depths. If only features from a certain depth are supposed to be recorded (e.g., for pseudoslices), an array should provide a pronounced response for a limited depth range. If, however, the aim is to detect all buried features with a single measurement regardless of their depth, a more integrated response is desirable.
- ***Ease of interpretation***: As was shown above, the anomalies produced by a simple buried sphere can be very complex. To aid with the interpretation of anomalies their shape should be as simple as possible.
- ***Ease of use***: As a practical aspect, it is required that an electrode array for archaeological prospecting can be used easily, and measurements over considerable areas can be conducted in reasonable times.

When comparing Wenner (see figure 3.8) and twin-probe arrays (see figure 3.9), it is clear that the resistivity response of the former is greater. This can be explained by the weaker current density of the twin-probe array, as the injected current is dispersed up to the distant remote electrodes. When investigating the spatial resolution, though, a twin-probe array produces considerably tighter anomalies that are indeed not much wider than the buried feature. This is mainly due to the smaller size of the active (i.e., mobile) electrodes: the array's overall width is a for the twin-probe array compared to $3a$ for the Wenner array. It can also be seen that for a Wenner array the strength of anomalies changes considerably with the depth of a feature, whereas smaller variations are visible for twin-probe arrays when comparing traces from different depths. A Wenner array is therefore better suited for pseudosections where a good depth resolution is needed, while a twin-probe array is superior at recording all features at different depths with similar amplitudes.

Anomalies produced by both Wenner and twin-probe arrays can have very peculiar shapes even over simple features (e.g., a buried sphere). In most cases, the Wenner response will maintain its overall shape, maybe with small peaks to both sides of the central anomaly. For the twin-probe array, however, the anomalies can develop into pronounced peaks to the side and an inversion directly over the buried feature (see figure 3.10). As was shown for the data from Ripley (section 3.4, figure 3.11), an awareness of such possible complications is required for the correct interpretation of results. It also means that the separation of electrodes should be adjusted according to the size of expected features. Good results are obtained for $A = 1$, which means that the separation of the array's mobile electrodes should be about half the size of the archaeological targets. The conventional use of a 0.5 m separation is a good compromise between depth penetration (see Depth of Investigation, section 4.1) and reasonably simple shapes of the anomalies. However, for very narrow targets at shallow depth, such as single stones or walls close to the surface, peculiar anomalies may still be produced.

To achieve a certain electrode separation a, the overall width of a Wenner array has to be $3a$, whereas the mobile electrodes of a twin-probe array will only be separated by a, as noted before. It is obvious that moving these two electrodes around a survey area is much easier than moving the four electrodes of a Wenner array—despite the need for a long cable to connect mobile and remote electrodes in a twin-probe array.

On balance, the twin-probe array shows some clear advantages for archaeological prospecting and has been established as the most commonly used array since its first discussion in 1968 (Aspinall and Lynam 1968; 1970).

5.2 Survey procedure

The fundamental considerations about earth resistance measurements from previous chapters are also useful to optimize survey procedures in the field.

5.2.1 Survey directions

As will be discussed in section 6.1, archaeological geophysical surveys are normally carried out over a regular grid, with readings taken at certain intervals along the grid's traverses. An earth resistance array will usually be kept in the same alignment relative to these traverses, and most surveyors prefer an arrangement in which the line of electrodes is perpendicular to the traverse.[3] In this case the earth resistance readings recorded along such a traverse follow the pattern of a transverse traverse and are therefore showing a simpler anomaly shape, as mentioned in section 3.4. Since data are recorded on a regular grid, a set of measurements can be extracted from the collected data that is perpendicular to the traverses' direction. When looking at such a perpendicular line the sequence of resistance values correspond to a longitudinal traverse (figure 5.2), and therefore it shows a more complex shape.

The effect can be seen in the data for the aforementioned graveyard at Ripley (sections 3.4 and 4.2.2; figure 3.11), where the survey direction was from left to right (west to east). Along the horizontal lines (transverse transect) the response is simpler than in the vertical direction (north-south, longitudinal transect). It can be concluded that if the alignment of shallow and narrow features is known, it is best to align the survey direction perpendicular across them, if no other factors determine the grid layout. In this way the measured anomalies will be easier to interpret.

5.2.2 Remote electrodes

When undertaking twin-probe earth resistance surveys the remote electrodes are usually considered to be an infinite distance away from the mobile electrodes and therefore irrelevant for the survey results. In practice, however, this distance is finite and variable, and the effects of this on the measurements will be discussed in the following section.

Distance

In an area survey, usually the earth resistance data are mapped as the final result, not the apparent resistivity. It is therefore necessary to consider

---

3. This is not a necessity, however, and members of English Heritage's Ancient Monuments Laboratory were known to align the electrode array with the traverse (Clark 1996, figure 35).

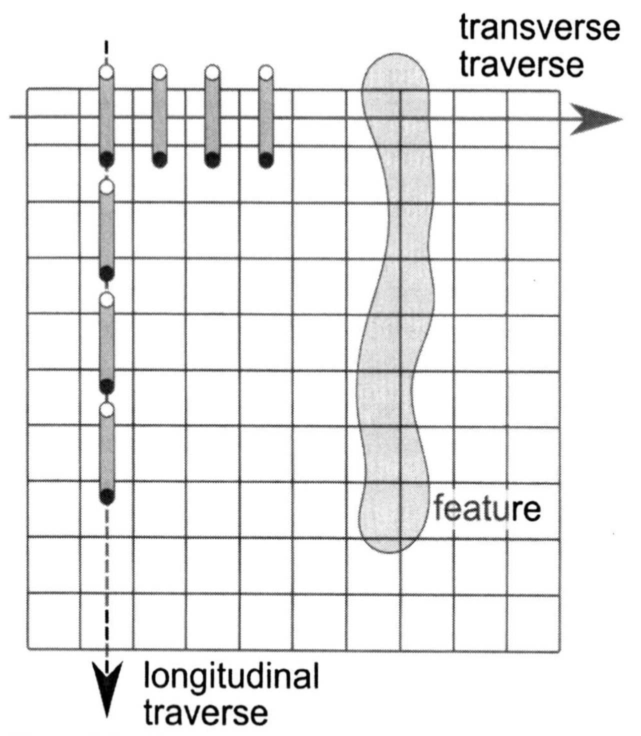

Figure 5.2. From a two-dimensional grid of measurements linear transects of data can be extracted that resemble either longitudinal or transverse traverses.

how the resistance of a finite twin-probe array changes if the mobile electrodes are moved relative to the remote electrodes. To assess this situation the simplified finite twin-probe array of figure 2.13 is considered ($b = a$). The relationship between measured resistance and the resistivity of a homogeneous halfspace is given in equation 2.18. When compared with equation 3.1, the array index is calculated as

$$n = \frac{C+1}{2C} = \frac{1}{2}\left(1 + \frac{1}{C}\right). \tag{5.1}$$

This array index can describe both a Wenner array ($C = 1$, resulting in $n_W = 1$) and an ideal twin-probe array ($C = \infty$, yielding $n_T = 1/2$). The question, then, is how much this index (and hence the measured resistance) changes if the distance C varies. This dependency on C is estimated by the first derivative of n as

$$\Delta n \approx \left|\frac{\partial n}{\partial C}\right| \Delta C = \frac{1}{2C^2}\Delta C, \tag{5.2}$$

where Δn and ΔC are the changes in n and C, respectively. From this relationship the relative, or percentage, changes can be calculated as

$$\frac{\Delta n}{n} = \frac{\Delta C}{2C^2}\cdot\frac{2C}{C+1} = \frac{1}{C+1}\cdot\frac{\Delta C}{C}. \tag{5.3}$$

To assess the variation in earth resistance, its relative change can then be related to the array index (see equation 3.1) as $\Delta R / R = \Delta n / n$. An example may help to illustrate these findings. Given a twin-probe array with a mobile electrode separation of $a = 0.5$ m and a distance between the two electrode pairs of $c = C\,a$, it shall be considered how much the earth resistance changes when moving the mobile electrodes from one grid position to the next. If the grid interval is 1 m, then such a step ($\Delta c = 1$ m) would mean a change in C by 2 ($\Delta C = 2$). Based on equation 5.3, the resulting change of measured earth resistance over a homogeneous halfspace is listed in table 5.1 for different distances between the electrode pairs.

Even for the homogeneous ground, without any buried features, the measured earth resistance varies considerably as the twin-probe array is moved across the measurement grid. This effect is caused by the varying geometry of the overall electrode arrangement. It can be minimized if the remote electrodes are a considerable distance away (e.g., $C \geq 30$).

Table 5.1. Change in earth resistance when moving a 0.5 m twin-probe array from one grid position to the next ($\Delta c = 1$ m)

Distance to remote electrodes (c)	*C*	*ΔR / R*
2 m	4	10%
5 m	10	2%
10 m	20	0.5%
15 m	30	0.2%
30 m	60	5×10^{-4}

A similar insight can be obtained when approaching this issue in a slightly different way. As discussed before, a twin-probe array is never used in its ideal configuration where the remote electrodes are an infinite distance away ($C = \infty$, $n_T = 1/2$). The relative deviation of the real array index from its ideal value can be calculated from equation 5.1 as

$$\frac{n - n_T}{n_T} = \frac{n}{n_T} - 1 = \frac{1}{C}, \tag{5.4}$$

and the measured earth resistance will deviate from the ideal value accordingly. The important observation is not that there is a deviation (as this could easily be corrected), but that it *varies* across a site as the distance between the electrodes changes. Typical values of the deviation are listed for a 0.5 m twin-probe array in table 5.2. The variation with changing distance is very apparent.

To minimize this effect, two possible solutions are conceivable. One is to chart the correctly calculated apparent resistivity, not the measured earth resistance, taking the exact distance to the remote electrodes (C) into account. However, this would require an accurate recording of the remote electrodes' position, which can be cumbersome and may not be feasible. The other possibility is to use a large distance between the remote and mobile electrodes such that changes in earth resistance due to the varying electrode geometry are negligible. This is the most commonly adopted approach. Aspinall and Lynam (1970) suggested that a relative deviation of 3 percent is acceptable and that the distance to the remote electrodes should therefore be $C = 30$; that is, $c = 15$ m for a 0.5 m twin-probe array (see table 5.2). This arbitrarily chosen value has proven suf-

Table 5.2. Deviation of real earth resistance measurement from an ideal 0.5 m twin-probe array

Distance to remote electrodes (c)	*C*	*Relative deviation*
2 m	4	25%
5 m	10	10%
10 m	20	5%
15 m	30	3%
30 m	60	2%

ficient for practical purposes and has been recommended ever since—for example, by Clark (1996).

Separation

Even with a long enough cable to minimize the effect of a varying distance toward the remote electrodes, it will normally not be possible to use the same position of remote electrodes for the whole survey area. When moving the remote electrodes to accommodate this, a different ground resistivity at the new location will lead to slightly changed earth resistance readings. In this case adjacent measurements that were recorded with different positions of remote electrodes may differ, leading to undesirable "steps" in the resulting data blocks. A common approach to minimize this effect is to adjust the separation between the two remote electrodes at the new location until the same reading is obtained as at the initial location (Gaffney and Gater 2003).

To study this procedure reference can be made to the ideal twin-probe array with different separations of mobile and remote electrodes, a and b, respectively (see figure 2.12). Earth resistance R and array index n for this configuration are calculated using equation 2.16 and equation 3.1,

$$R = \frac{\rho}{2\pi a} \frac{1}{n} \text{ with } \frac{1}{n} = 1 + \frac{1}{B} \tag{5.5}$$

where[4] $B = b/a$. It is therefore clear that by changing the separation of the remote electrodes the measured earth resistance will change as well. Equation 5.5 shows that this change is not in the form of an offset, but all resistivity values are scaled with a factor related to $1/B$. This implies that the overall range of measured earth resistance values is also changed, thereby altering the data's contrast. As an example, the resistivities of the ground may be assumed to vary between 150 Ωm and 200 Ωm. For an ideal twin-probe array with a mobile electrode separation of a = 0.5 m, table 5.3 lists the expected earth resistance readings at different separations between the remote electrodes and the difference between these values as the "range" of

4. Often B is considered to be nearly 1, resulting in n = 1/2; see section 2.4.1.

Table 5.3. Dependency of the earth resistance range on the remote electrode separation *b* for ground resistivities between 150 Ωm and 200 Ωm and a twin-probe array with mobile electrode separation of *a* = 0.5 m

b [m]	*B*	*R [Ω]*		*Range [Ω]*
1.5	3	63.7	84.9	21.2
2.0	4	59.7	79.6	19.9
2.5	5	57.3	76.4	19.1
3.0	6	55.7	74.3	18.6

recorded readings. It can be seen that the range, or contrast, decreases with increasing separation between the remote electrodes.

To quantify the relative change of earth resistance with changing separation of the remote electrodes, it is necessary to form the first derivative of the resistance R with respect to the relative separation B:[5]

$$\frac{\Delta R}{R}=\frac{1}{R}\frac{\Delta R}{\Delta B}\Delta B\approx\frac{1}{R}\left|\frac{\partial R}{\partial B}\right|\Delta B=\frac{1}{B+1}\cdot\frac{\Delta B}{B}. \tag{5.6}$$

Taking as an example the 0.5 m twin-probe array and an initial remote electrode separation of 2 m (i.e., B = 4), an expansion of that separation by 0.5 m (ΔB = 1) will result in a notable relative earth resistance change of $\Delta R / R$ = 5%, according to equation 5.6.

It follows from this discussion that the method of balancing data grids by changing the remote electrodes' separation can achieve comparable values in adjacent data grids but that the overall contrast may be different for the individual locations of remote electrodes, possibly leading to a patchy appearance when plotting the data. To correct for this effect, multiplicative scaling would have to be applied to the data values, which is cumbersome to achieve with currently available software. It may be easier to keep the remote electrodes at a fixed separation throughout the whole survey (e.g., by fixing a piece of string between them) and apply slight offset corrections to the resulting data later using the standard software tools, such as edge matching (see section 6.2.1). At least the data's contrast can then be maintained throughout the survey. However, equation

5. The derivative is $\frac{\partial R}{\partial B}=-\frac{\rho}{2\pi a}\frac{1}{B^2}$.

5.6 suggests another possible option. If B is chosen to be very large (e.g., $B = 20$), the relative resistance changes become negligible. The disadvantage of this method is the overall reduction of measured earth resistance to about half the usual value ($n \approx 1$ instead of $n_T = 1/2$). Some sensitivity to small resistivity variations may therefore be lost, but with modern instrumentation this will be hardly noticeable. Bevan (2000) arrives at the same conclusion using a slightly different argument when discussing the pole-pole electrode array. It is, however, clear that none of these different configurations are "wrong" in the sense of producing unusable data; it is just that some may make subsequent data improvements easier.

Orientation

It has been shown above that readings recorded with a real twin-probe array may be influenced by the survey procedure. Related to this is the possible effect of the electrode orientation. When surveying an area along parallel traverses, it is most efficient to record measurements in adjacent lines by walking in opposite directions (i.e., walking zigzag). It is often the case that the electrode array is rotated at the end of a line so that the operator maintains the same relative orientation to it (figure 5.3a). However, this means that the orientation of the two mobile electrodes relative to the remote electrodes is reversed, which may lead to small variations in the recorded earth resistance due to the different geometry. This effect will only be discernible as faint stripes in the recorded data, if the distance to the remote electrodes is not "sufficiently" big (e.g., $C < 30$). If there are any doubts about the influence of this effect, the rotation of the mobile electrodes between adjacent lines should be avoided (figure 5.3b).

5.3 Other factors affecting earth resistance measurements

5.3.1 Environmental influences

Due to the electrolytic nature of earth currents, ground resistivity depends crucially on moisture content (see section 1.3.2). Inevitably, environmental factors (e.g., rain, wind, sunshine) can have pronounced effects on earth resistance measurements. It is hence useful to discuss them in relation to overall site properties and to the resistivity contrast of individual features.

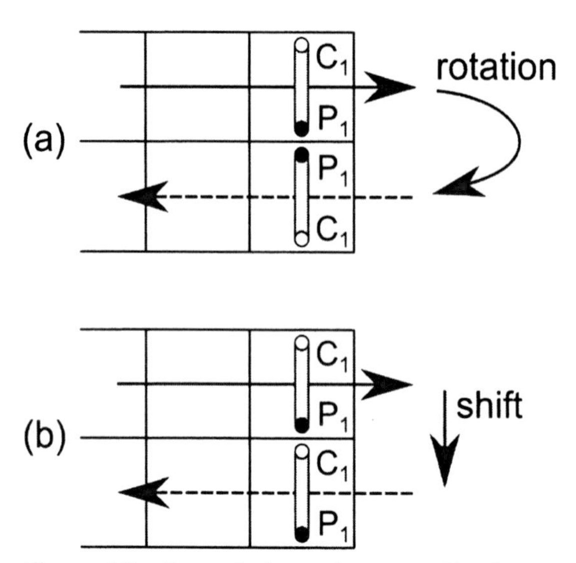

Figure 5.3. For a twin-probe array the frame with the mobile electrodes can be moved in two different ways at the end of a line: (a) it can be rotated, which changes the overall electrode layout; (b) it can be shifted, leading to a more consistent geometry.

Effects on a site

When the soil moisture content of a site changes (e.g., due to rain), the range of resistivity values will be not only shifted but also scaled. For example, if the resistivity of a dry site varies spatially between 150 Ωm and 200 Ωm, after prolonged rain the same site may have resistivities between, say, 20 Ωm and 30 Ωm; the site's overall resistivity contrast has changed. Therefore, when recording earth resistance data over different days, seasons, or even years, the data sets may have diverse ranges that are very difficult to match into an overall display. Simple grid balancing that adds different offsets to the various data areas will be insufficient, and more complex normalization procedures are required that allow a scaling of results (see section 6.2.1).

Resistivity contrast of features

Buried archaeological features can only be detected with earth resistance measurements if they exhibit a resistivity contrast to their surrounding background material. Depending on whether this contrast is positive or negative, the features will produce anomalies that are higher or lower than the background level (positive or negative anomalies, respectively). It was already discussed in section 1.4 how weather and geological conditions can influence the resistivity contrast of a ditch. Generally it can be said that prolonged very dry or very wet weather produces poor contrast and that the transition between wet and dry weather facilitates the detection of features. Not only the input of rain has to be considered when assessing environmental influences, but also the loss of water. The related mechanisms are summarized as evapotranspiration that includes the effects of wind, sunshine, or variations in vegetation. While an important parameter, evapotranspiration is difficult to quantify or predict. It is an important aspect of studies in soil science (Novak 2012).

In a study by Cott (1997) earth resistance measurements were conducted at the Roman town of Caistor (Norwich, United Kingdom; see section 3.5.2) over a thirteen-month period. The results over one particular ditch were monitored, and the corresponding sequence of twin-probe pseudosections is shown in figure 5.4. As explained earlier (section 4.2.1), the shape of the pseudosection anomaly does not fully reflect the presumed triangular shape of a Roman ditch, but the variation of the resistivity contrast over time is quite apparent. The study found that there is a lag of several weeks between the input of rain and the corresponding change of the earth resistance anomalies. The inverse resistivity range, which describes the "narrowness" of the range of apparent resistivity values, followed closely the total amount of monthly rainfall (figure 5.5). For example, in September 1995 a total of 80 mm rainfall had been recorded, and the pseudosection for that month showed a particularly narrow range of apparent resistivity values (from 261 to 359 Ωm). In other words, when it was particularly wet the apparent resistivity values varied little across the pseudosection.[6] Nevertheless, this variation was sufficient to discern the anomaly that was created by the ditch. This highlights that the absolute value of apparent resistivity in a pseudosection is less important than the contrast between a feature's anomaly and the background values.

6. The inverse resistivity range was better correlated with the monthly rainfall than the mean of the apparent resistivity values.

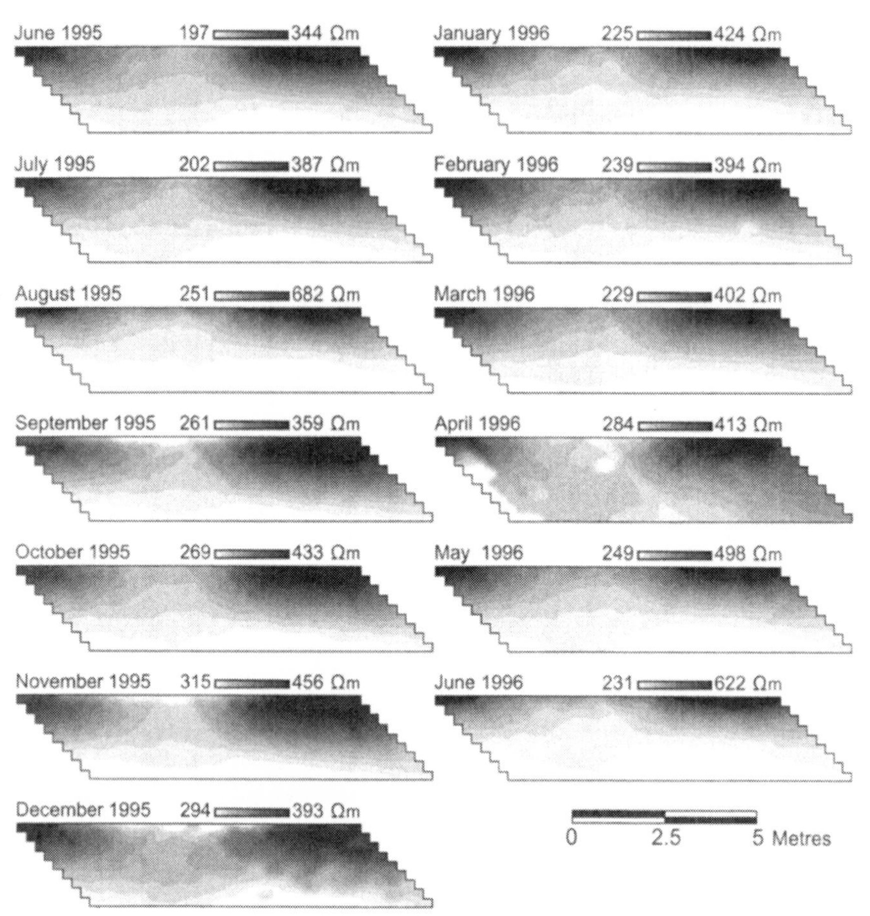

Figure 5.4. Twin-probe ERI pseudosections over a Roman ditch at Caistor, monitored over a thirteen-month period.

5.3.2 Instrumental problems

Problems caused by the measuring instruments are also often related to environmental factors. In very dry conditions, for example, the contact of electrodes with the ground can be very poor (high contact resistance), and although a resistance reading is logged it may be just a singular high value or so-called spike. In such conditions it may be better to enter a blank reading ("dummy reading") rather than forcing electrodes in the ground to establish a poor current path.

By contrast, wet conditions sometimes allow currents to flow outside the ground. For example, wet grass may form a shortcut between some

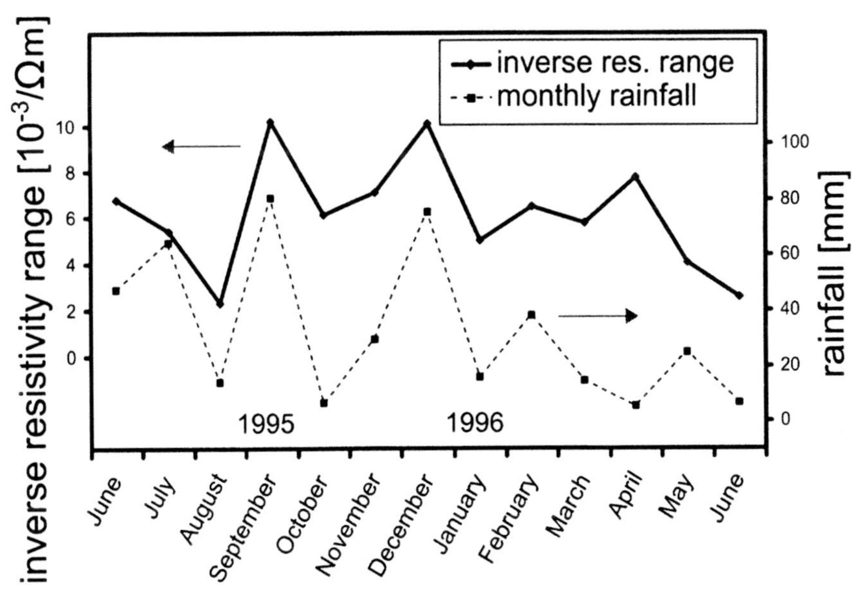

Figure 5.5. For the ERI data from the Roman ditch at Caistor the inverse resistivity range (i.e., the "narrowness" of the data range) follows the monthly rainfall pattern very closely.

of the electrodes, hence causing erroneous readings. Similarly, a current path can be established across a film of water on clothing and the wet hands of an operator. In both cases the measured resistance will only have a vague relationship to the features in the ground. These effects can be minimized by insulating as well as possible all parts of the earth resistance array against each other (e.g., plastic washers, wooden or plastic beams for the electrodes).

A trivial, but sometimes overlooked, cause of poor measurements can be a loose contact between electrodes and the earth resistance meter. Readings, albeit erratic, can sometimes even be recorded if an electrode is plugged into an unused connector. Small voltages induced in the neighboring cables may be enough to create a false reading.

5.3.3 Topographical effects

It is often observed that existing surface features (e.g., banks) cause earth resistance anomalies. Two possible causes have to be distinguished: differential drainage and the resistance topography effect.

Differential drainage is most pronounced on the flanks of topographic features—for example, on the slopes of banks of earth monuments. The exact details strongly depend on the specific soil conditions on a site. In most cases the slopes of a bank will drain faster than the top due to gravitational forces, leading to raised resistivity along the flanks. However, if the erosion of an earth monument has led to the deposition of considerable quantities of soil on the slopes, they may retain more moisture and therefore have lower resistivity than the top. For the flanks of a wide ditch, which still shows as a pronounced surface depression, similar considerations apply. For such a ditch the bottom usually acts as a sink where soil and moisture accumulate, and it often has a low resistivity. For the interpretation of earth resistance data over topographic features it is important to know the exact location of slopes and depressions, and it is useful to record these during the time of the geophysical survey—for example, in the form of a simplified earthwork map. Whether the soil on a slope has low or high resistivity will also depend strongly on a site's past weather conditions, and an awareness of these issues is required when interpreting the data.

At the southern Thornborough henge (North Yorkshire, United Kingdom), the circular bank survives to a height of about 1.5 to 2 m (figure 5.6). On its top (north part of the image) the earth resistance values were high (180...200 Ω), decreased southward down the slope (140...160 Ω), and increased again right at the bottom of the slope (180...230 Ω) where the soil was well drained (Schmidt et al. 2006). Further south, over the inner ditch, the earth resistance dropped to 30 Ω. This ditch was only visible as a very shallow surface depression, and the low readings were caused by the moisture retention of its fill. However, the bank and its slope had a pronounced topographic effect on the soil's resistivity and the measured earth resistance.

As was discussed in section 4.2.1, abrupt changes of topography can introduce anomalies even in earth resistance data over homogeneous ground (Fox et al. 1980; Yilmaz and Coskun 2011). However, it was shown in a study by Sutherland et al. (1998) that these effects are normally negligible for archaeological features because slopes over soil-covered archaeological monuments are usually fairly smooth (i.e., without sharp kinks) and the span of electrode arrays often only covers a single aspect of a slope (i.e., not up the slope, across the top and down again). Neverthe-

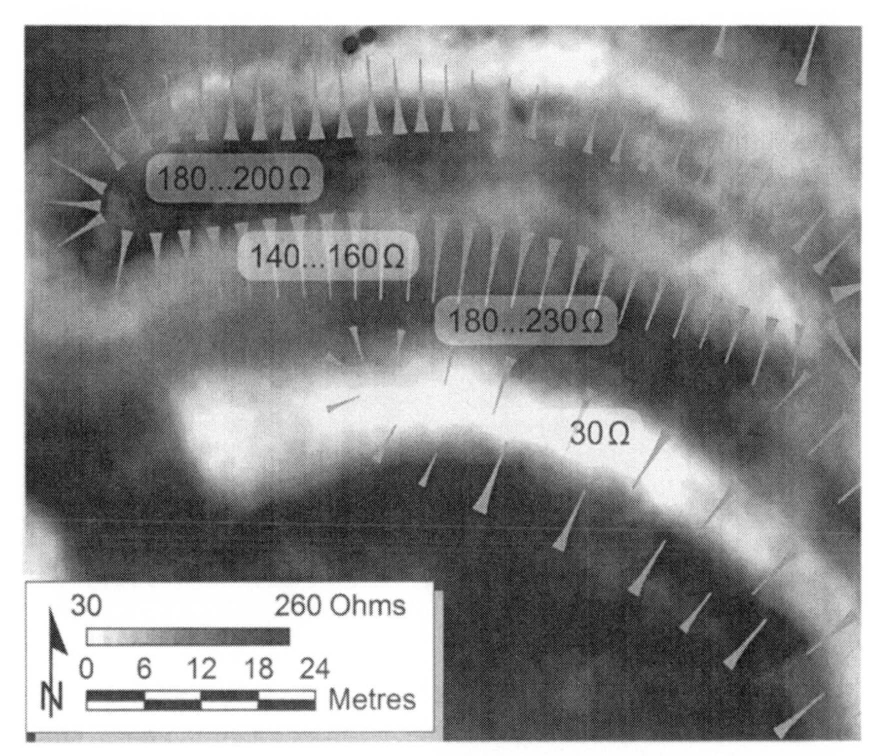

Figure 5.6. Northeastern bank of the southern Thornborough henge. The earth resistance along the slopes is lower than at the top or bottom due to moisture retention in the soil of the flanks.

less, it is important to record topographic variations whenever there is a possibility that they may introduce artifacts in the data so that this can be taken into account when interpreting the results. Where anomalies are found on the slopes of earthwork features they are unlikely to be caused by structures underneath the slopes. Anomalies that reach beyond the slopes are more likely caused by archaeological features.

CHAPTER SIX

DATA TREATMENT

There are three main reasons why the computer manipulation of acquired field data may be desirable or even necessary. First, the data may have "defects" due to problems during data acquisition. These can be either issues related to the equipment (e.g., erroneous readings), errors of the operator, or the environmental conditions encountered during the survey. Second, the data may require processing. For example, earth resistance data collected for pseudosections have to be converted into apparent resistivities, and tomography data have to be tomographically inverted. It may also be the case that the features in the ground only produce very weak anomalies in the data, so that filters have to be applied to highlight these anomalies. And third, it is often useful to present the data in an easily understandable form to the intended audience (e.g., geophysicists, archaeologists, site managers, building developers, general public), and therefore some image enhancement may be required. This chapter will discuss these three categories of data manipulation techniques in turn. They can be summarized as data improvement, data processing, and image processing (Schmidt 2003; Schmidt and Ernenwein 2011). Some of the filtering and visualization approaches are similar to those used for magnetometer data and are discussed in detail elsewhere (e.g., Aspinall et al. 2008).

As a general rule, it is advisable to manipulate data as little as possible as each extra step has the potential to either remove information or introduce processing artifacts that may then falsely be interpreted as archaeological features. Some practitioners therefore also distinguish between raw data (as collected in the field), minimally improved data (minimal manipulation that does not introduce artifacts, but allows seeing all data values in a single plot), and processed data (everything else). In this classification minimally improved data are deemed "just about acceptable," as they allow seeing the nearly raw data (for example, by balancing grids; see section 6.2.1), and any further choice of data improvement and processing steps is deemed a subjective alteration of the data and their information content. This is similar to the discussion of ERI inversion schemes where, in the absence of other criteria, sometimes simply the "best looking" inversion result is chosen (see section 4.3), which implies that those processing the data already "know" what they want to see. It also holds for general data treatment that there is no "correct" processing flow and that processed data are not "better" than minimally improved or raw data. If an experienced archaeological geophysicist cannot see hints of anomalies in the minimally improved data, then interpreted features should not normally be derived from any of the more advanced processing steps, as they may turn out to be processing artifacts.

6.1 Data assembly

Most manipulation techniques for measurements from earth resistance area surveys rely on gridded data. This means that all data are assembled in a regular raster, like pixels in a digital photograph. The raster can then be described by its resolution in two perpendicular directions, usually referred to as *x*- and *y*-directions, or easting and northing. For example, a twin-probe survey undertaken along traverses running west-east that are spaced 1 m apart, and where readings were collected every 0.5 m along these traverses, would have an *x*-resolution of 0.5 m and a *y*-resolution of 1 m. It is a common convention to choose the *x*-direction as being along the line of traverses, regardless of its orientation relative to north.

In most cases data for geophysical surveys are also collected in the field in such a raster. For earth resistance measurements, large survey areas are frequently split into smaller chunks—for example, of 20 m × 20 m size. These partitioned areas are referred to by different names; for example, by archaeological geophysics instrument manufacturers simply as "grids," by remote sensing experts as "tiles," and sometimes as "subgrids."

In accordance with Aspinall et al. (2008) the term "data grids" is used here. They allow subdividing a large survey area into manageable sections, which can then be indexed through row and column counts (Schmidt and Ernenwein 2011). Twin-probe earth resistance measurements require the relocation of the remote electrodes when the cable length becomes insufficient for more distant recording positions. Undertaking such a survey using data grids makes these operations far more efficient. It should also not be underestimated that subdividing a large area into individual data grids greatly helps with the motivation of tired survey personnel ("only two more grids to go"). It is important to record the sequence in which they are being investigated during a survey—for example, to identify all data grids that were measured with the same remote electrodes. Obviously, it is also essential to provide clear indications as to where each data grid is located in the overall geophysics grid for the whole area. This can be done either through a map of the data grids, visualized as a mesh (also referred to as a "master grid"), or by specifying the *x* and *y* coordinates of the corners of each data grid. Using this information the measurements from all data grids can be assembled into one large data structure for the whole geophysics grid, often referred to as a "composite" (i.e., data grids + mesh = composite; see figure 6.1). The composite will also take the form of a gridded data structure, but for a much larger area. It allows data processing irrespective of the arbitrary data grid boundaries that were introduced during the survey.

One particular aspect of archaeological geophysics data is the use of "dummy readings," which are specific data values that are inserted into data grids and composites in order to indicate locations where no actual measurement was collected. This could, for example, be where a data grid cannot be filled entirely due to the proximity to a wall or field boundary, or when there was an obstacle along a traverse, like a tree, that prevented a measurement. These "dummy readings" are usually expressed as data values that are not normally used (e.g., 2047.5 or -999) or are given the specific representation of NaN (not-a-number) according to the IEEE 754 standard. In either case it is essential that the data-manipulation software can handle these special markers and does not use their numerical representation for any calculations.

Earth resistance data collected with wheeled arrays (see section 2.4.3) can accurately record the location of each measurement with GPS, and

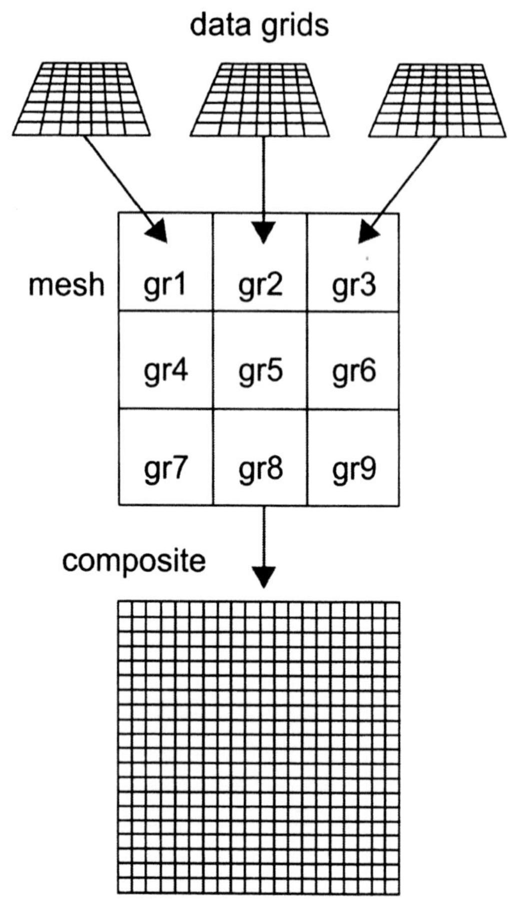

Figure 6.1. Using several data grids and a mesh that indicates their relative position, data can be combined into a composite.

the resulting data are no longer confined to a regular grid. There are some basic data improvement steps that can be applied to these data (e.g., spike removal along traverses, even if not entirely straight), but for any other data processing the data have to be "gridded" first to be available as a regular raster. The shortcomings of individual gridding algorithms (Li and Götze 1999) should be carefully considered, and the mechanisms for filling gaps between data tracks must be taken into account when interpreting the results.

6.2 Data improvement

6.2.1 Grid balancing

For twin-probe area surveys the remote electrodes may have to be moved between data grids, and sometimes surveys are undertaken on different days, or even in separate field campaigns. As a result, there can be differences between the data grids that need to be evened out before they can be treated as a seamless data composite. This is often referred to as grid balancing, and three issues need to be considered.

- *In-grid variation.* There may be gradual variations of readings within a data grid dependent on the electrode configuration. In section 5.2.2 it was discussed how twin-probe arrays show changes in the earth resistance readings even over homogeneous ground if the distance to the remote electrodes is too short. Similar effects are observed with some other arrays—for example, the Schlumberger array.
- *Offset.* If surveys of adjacent blocks were undertaken in different weather conditions, there may be differences in overall background resistance (e.g., high when dry, low when wet).
- *Contrast.* In most cases changes in environmental conditions also affect the range of possible values, often referred to as the contrast. For example, in dry conditions the data range may be 100...200 Ω, but the same site may exhibit a range of readings between only 20...70 Ω when wet. Hence not only the offset but also the contrast has changed. Changing the remote electrode separation in a twin-probe array also leads to a change in contrast as discussed in section 5.2.2.

In-grid variations can be very difficult to correct. Usually the only way is to record precise information about the location of all electrodes during the survey (including the remote electrodes) and to calculate the resulting apparent resistivity of each data point exactly according to its position within the data grid. It is therefore strongly recommended to minimize any such problems by avoiding electrode layouts that enhance this effect. For twin-probe surveys the usual recommendation is to maintain a distance to the remote electrodes of approximately thirty times the separation of the mobile electrodes (section 5.2.2).

Differences in offset and contrast often occur together, and it is usually best to treat them simultaneously. However, this is not implemented in conventional data processing software packages, which mostly concentrate on the adjustment of offsets alone. It can therefore be argued that it is important to maintain the same contrast when undertaking surveys over a large area to use the data processing facilities of available software most effectively. For twin-probe arrays this would mean that the same separation of remote electrodes is used throughout a survey. When they have to be moved to a different location to accommodate the available length of cable, the new block of data may have a different offset, but at least the geometrically related contrast will be the same. This procedure is different from the recommendations by Clark (1996: 46) and Gaffney and Gater (2003: 33), who suggest adjustment of the remote electrode separation to make readings of adjacent survey blocks similar, even though this may change the data contrast.

To adjust the offset between neighboring data grids, the procedure of edge matching is best suited. Data on both sides of the edge between neighboring data grids are automatically evaluated, and the offset of one grid is then adjusted. This can be done by choosing the edge of a particular grid and automatically adjusting one neighbor's offset, by manually adjusting the offset with a slider bar in some software packages, or by simultaneously calculating offset values that minimize *all* edge mismatches in a dataset (Haigh 1992). The Zero Mean Grid method that is often used for leveling magnetometer data (Aspinall et al. 2008) is normally unsuitable for earth resistance data, as these often exhibit a gradually varying background trend.

Adjusting the contrast between different data grids requires a few more processing steps, as the measurement values have to be scaled through multiplications. The process is best illustrated with reference to the data's histogram. Figure 6.2a shows the histograms for two example data blocks that could have been collected in wet (20...70 Ω) and dry (100...200 Ω) conditions, respectively. The histograms show on the *y*-axis the number of readings for a particular data value, and in this example both data blocks contain 1,600 readings. However, as these are spread over 50 Ω and 100 Ω, respectively, the histogram peak for the wet condition is twice as high in

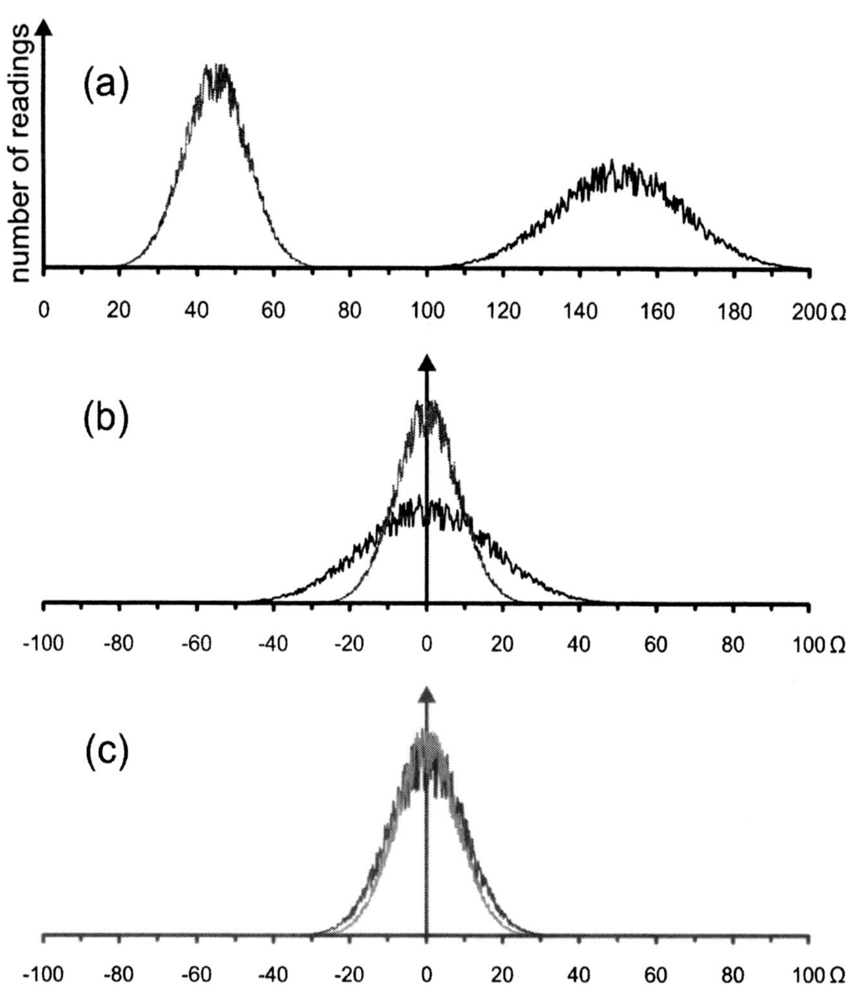

Figure 6.2. The artificial histograms of two survey blocks can be adjusted by shifting and scaling the data values: (a) the data collected in wet and dry conditions show as distinct peaks on the histogram (20...70 Ω and 100...200 Ω, respectively); (b) by subtracting the middle value from each peak both can be shifted to the center; (c) by subsequently multiplying each data block with an appropriate scale factor they can be made very similar.

this example. Figure 6.3a shows a data plot with similarly clear changes of contrast between the different survey areas. These data from the southern Thornborough henge (North Yorkshire, United Kingdom) were collected with students in several separate field seasons, with different soil moisture content in each. The data grids in figure 6.3 were all automatically edge matched using the Contors software (see section 7.1).

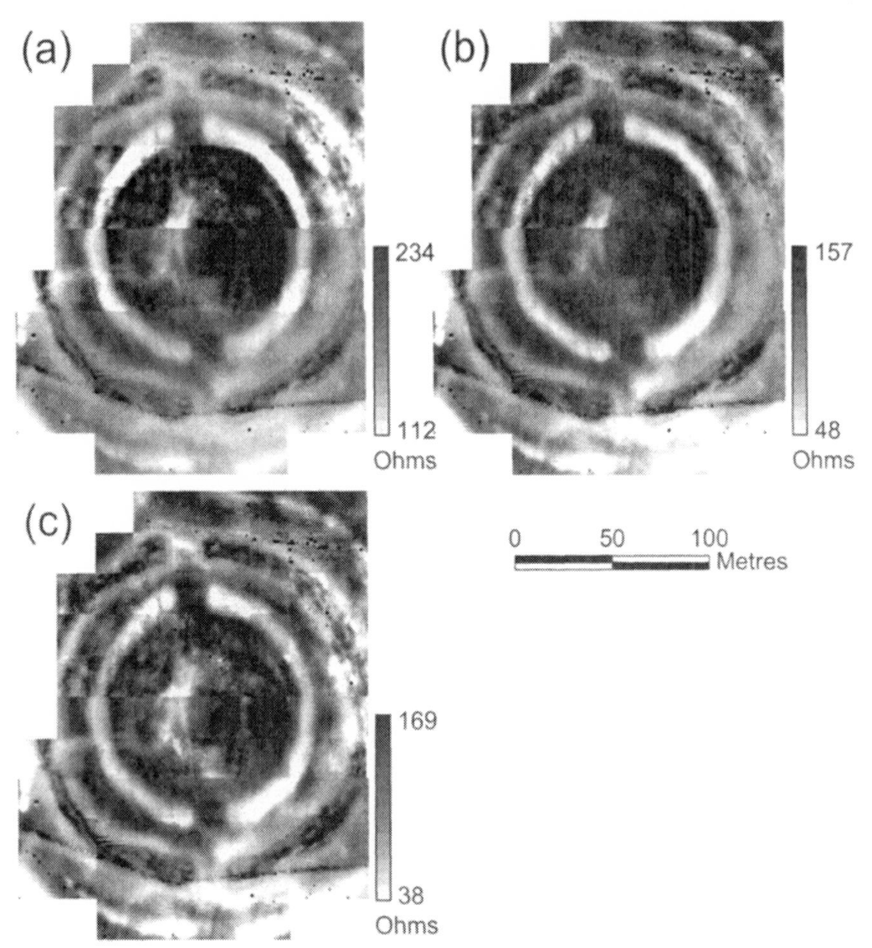

Figure 6.3. Twin-probe earth resistance data for the southern Thornborough henge were collected in several field seasons, leading to different contrasts in the data blocks: (a) original field data show clearly the different data blocks; (b) each data block was normalized and all blocks then combined; (c) each individual data grid was normalized and the data grids then combined. In all three diagrams the data grids were edge matched after combining into the composite.

To equalize the contrast between such data blocks, several steps have to be applied, and best results are achieved if all data grids that were recorded under the same conditions (e.g., same day, same weather, or same remote electrodes) are treated together as one block.

1. The histogram of each data block has to be centered on 0 by determining its "middle value" and then subtracting this from

all data values in the block (figure 6.2b). The simplest choice of a "middle value" is the mean, but using the median is more robust as it disregards outliers.

2. By multiplying each data block with an appropriate scale factor, all histograms will be stretched to the same width (figure 6.2c). For example, if the histogram of the dry block originally spanned from 100...200 Ω, after subtraction of the middle value in step 1 it will range from -50...+50 Ω. By multiplying all data values in that block with a factor of 0.5 the histogram will then stretch from -25...+25 Ω, same as the wet data block in this example. To determine the data spread the easiest calculation is to use the standard deviation, but the 5...95 percentile range would be better, as it discards outliers.

3. As it would be peculiar to see negative earth resistance values in a data plot, as produced in step 1 (e.g., -50...+50 Ω, and then scaled to -25...+25 Ω in step 2), it is advisable to add a common offset value to all newly adjusted data to make them positive. The value chosen should reflect the original data. In the above example adding 45 Ω will bring all data to a range of 20...70 Ω.[1]

4. Although by now the different data blocks have been adjusted to the same contrast, there usually remains a step of data values between the adjusted blocks, as earth resistance data often show a trend across the survey area and the necessary adjustment of the "middle value" in step 1 disregards this (it was already mentioned that Zero Mean Grid is not a suitable data improvement step). To correct for this problem the step in the data values is first estimated by investigating several data values on both sides of the block boundary. This step value is then applied as an additive offset to the block with lower data values.

Figure 6.3b shows that this approach has greatly improved the differences in contrast between the data blocks of the data from the southern

1. As is clear from this example, if one data block is adjusted to the data spread of the other, the second data block does not have to be altered at all.

Thornborough henge. However, it is also clear that there remain differences, as the change of contrast varies between the different archaeological features. In this example this is most marked between the inner ditch (two semicircular, low earth-resistance anomalies) and the surrounding banks. Figure 6.3c illustrates that equalizing the contrast of all subgrids individually does not improve this situation either.

6.2.2 Spike removal

As discussed in section 5.1.2, earth resistance surveys are undertaken with four electrodes to minimize the effect of contact resistance. However, if the contact resistance is extremely large (as, for example, in sandy soils) or the current flow is impeded by an insulating object directly underneath an inserted electrode, the recorded earth resistance may be erroneous. An area survey over a very stony surface may hence show many isolated very high readings, which are often referred to as spikes. In an ERI survey such spikes may show in individual readings, or a whole set of measurements may be affected. For example, if one electrode suffers from excessive contact resistance, all measurements that used this electrode position may be too high, which can lead to a line of elevated data running diagonally down the plotted pseudosection (see, for example, the twin-probe pseudosection in figure 4.13).[2] It can be desirable to remove these unwanted effects and create "better looking" data, but it must be ensured that no archaeological information is eliminated in the process.

For area surveys spike removal is a procedure similar to low-pass filtering, where a filter window of a specified size is passed over the whole dataset. At each position the central value is checked and, if identified as a spike, adjusted. To do this the data values in the filter window are examined and their mean and standard deviation calculated. If the central value is found to be a certain multiple of the standard deviation higher than the mean of its neighbors, it is considered to be a spike. It can then be replaced with that calculated mean, producing a smooth overall appearance. Alternatively, it can be replaced with the threshold value, thus creating a marker for the location of the spike, but with far lower amplitude, hence with little effect on data statistics. Figure 6.4 shows twin-probe data from Tilaurakot (Nepal; see section 7.4), demonstrating that the resulting data appear

2. Such a line is only diagonal if a shifting center arrangement was used.

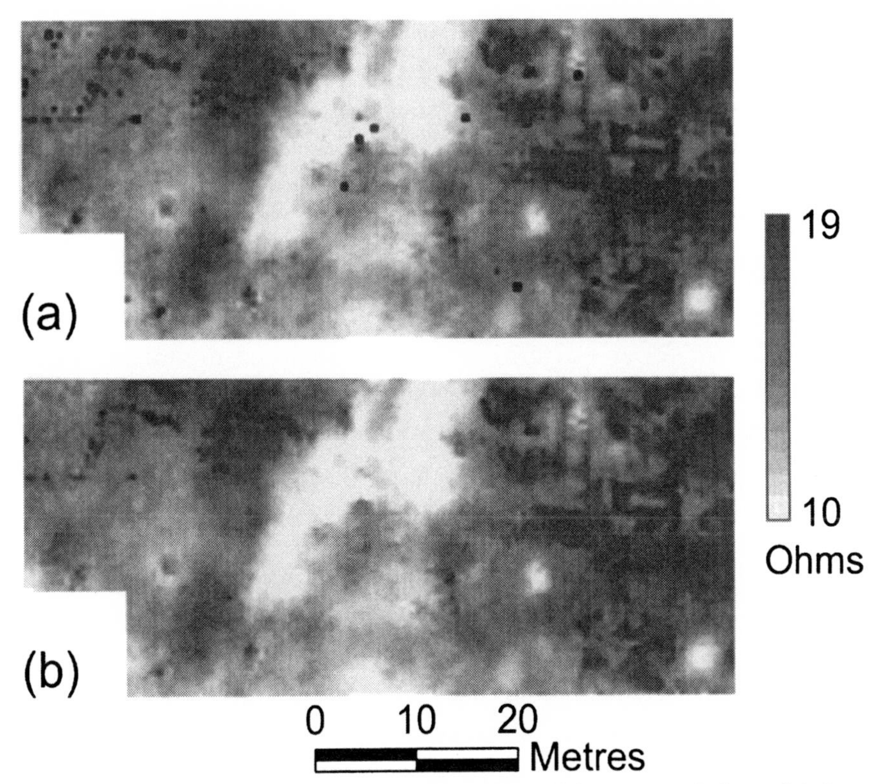

Figure 6.4. Twin-probe survey at Tilaurakot: (a) measured data; (b) data after spike removal, showing the rectangular structures more clearly.

smoother, and it is easier to focus the interpretation on the dominant rectilinear features. Another advantage of spike removal is that for software that automatically adjusts the display range of its data, the removal of outliers from the statistics allows for a much-improved display. ERI data are sometimes edited manually to minimize the effect of spikes, relying on user judgment as to which data points are deemed to be erroneous.

Sometimes, singular very high readings may in fact be real anomalies—for example, caused by a small feature of very high resistivity contrast (e.g., a small stone). In this case the anomaly and its side-wings may span several data grids. In such a case it is better to complete all other data improvement tasks first (especially grid balancing) and then treat such singular anomalies in the data processing stage. As with magnetometer surveys, assemblies of singular readings may have archaeological relevance, and the case for their removal has to be judged carefully. It may, for example, be that an area of particularly stony ground that produced data spikes is caused by the remains of a floor surface, or that small stone fragments were chipped off a buried stone feature (e.g., a Roman drain) by repeated plowing. In such a case the assemblage of these fragments may give an indication for the course of the feature.

6.2.3 Line adjustment

For magnetometer data the adjustment of adjacent survey lines with a Zero Mean Traverse algorithm is commonly used to ameliorate the defects that stem from imperfections of sensor balancing (Aspinall et al. 2008). By contrast, for earth resistance area surveys there are very few cases where such a data treatment may be useful. There have been situations where stripy data were produced when using a multiplexer to switch between different electrode configurations for one insertion point of a mobile electrode array. In particular, when two adjacent readings are being collected (e.g., left 0.5 m and right 0.5 m; see figure 6.5), there was found to be a slight offset in the data values. The reason for this is not entirely clear, but it is probably related to the timing of the individual readings collected with the multiplexer. As discussed in section 5.1.3, a delay can be set that determines how long after a first contact with the ground the measurement is recorded by the RM15. In contrast to this

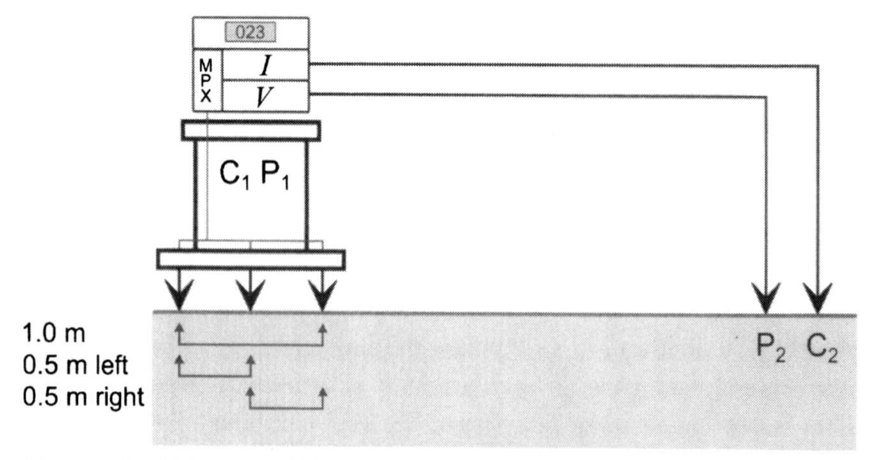

Figure 6.5. Using a mobile frame with three electrodes and a multiplexer allows one to collect three different twin-probe readings at each measurement position.

user-selected delay the timing for subsequent readings with the MPX15 multiplexer is determined by the switching circuitry. As a result, the delay for the first reading is different from the switching times of all subsequent measurements. The multiplexer should hence be programed in such a way that all readings that are supposed to be comparable (e.g., two adjacent readings with 0.5 m electrode separation) are taken with the same delay time. For example, the first reading for a three-electrode twin-probe configuration could be done with a 1 m electrode separation using the two outermost mobile electrodes, while the second and third readings (both with the delay from the multiplexer switch) use the left and right pair of the 0.5 m separated electrodes (see figure 6.5). Such a multiplexing sequence was shown to eliminate nearly all striping effects. Should a dataset be encountered with a consistent difference between readings of a multielectrode array, it may be necessary to experiment with a Zero Mean Traverse algorithm to improve the data. However, as with magnetometer results, care has to be taken not to remove relevant anomalies (e.g., walls) that run parallel with the survey lines.

As mentioned in section 5.2.2, rotating a twin-probe array's mobile electrodes while walking zigzag along parallel survey traverses can lead to changes in the measured earth resistance due to a change of the electrode array's overall geometry, showing as stripes in the data. However, this effect is usually only noticeable in surveys where the distance to the remote electrodes is far less than the recommended thirty times of the mobile electrode separation. In such a case the broad earth resistance variations that result from the changing distance to the remote electrodes will have a much stronger effect than any such stripes. The recommended distance to remote electrodes should therefore always be maintained.

6.3 Data processing

The most common data processing techniques are high- and low-pass filtering of gridded data. In their simplest form they remove variations in the data that are respectively larger or smaller than the archaeological features of interest. The original definition and realization of these filters in the frequency domain will be discussed after their more common implementation as kernel convolutions.

This convolution is achieved by using a small rectangular window (e.g., 3 × 3 readings in size) centered, in turn, over each data value to evaluate the background around the data point being processed and then multiplying the filter coefficients of this window (the "kernel") with the data values underneath, adding the individual terms and using the resulting value as a replacement for the central data point in question ("convolution"). This method is therefore called kernel convolution (figure 6.6).

Based on the choice of the filter coefficients, it can perform various filtering tasks. The size of the kernel has an influence on the results and needs to be chosen appropriately. For example, the size of a high-pass filter kernel determines the area that is taken into account when evaluating the data background (see below) and influences how smooth the resulting data will look. The size of the kernel is expressed either as the absolute number of coefficients in both directions (e.g., 5 × 3) or by specifying a "radius" from the center. A kernel of size 5 × 3 would have a radius of 2 in the x-direction and 1 in the y-direction. As kernels have to be symmetric about their center,[3] specifying a radius is a more natural choice

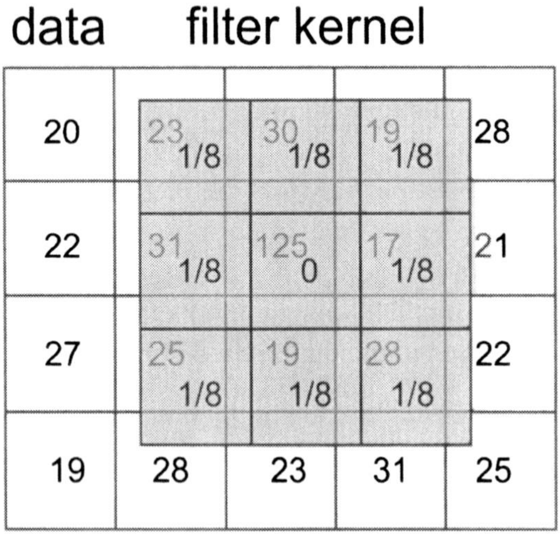

data filter kernel

20	23 1/8	30 1/8	19 1/8	28
22	31 1/8	125 0	17 1/8	21
27	25 1/8	19 1/8	28 1/8	22
19	28	23	31	25

Figure 6.6. The filter kernel is applied to each data position to calculate the filtered value. In this example the new value is 1/8 × 23 + 1/8 × 30 + 1/8 × 19 + 1/8 × 31 + 0 × 125 + 1/8 × 17 + 1/8 × 25 + 1/8 × 19 + 1/8 × 28 = 24.

3. A kernel of, for example, 2 × 4 is not possible.

for defining them. These sizes are expressed in data counts, not in real distances. If, for example, data were collected every 0.5 m along traverses (x-direction) that are 1 m apart (y-direction), the above-mentioned kernel of 5 × 3 would have an actual size of 2.5 m × 3 m, and would hence be as close to a square as can be achieved in this grid; a 3 × 3 kernel would result in a rectangular filter size of 1.5 m × 3 m. As the filter kernel is used to calculate a replacement for its central value, problems occur for data values directly at the edge of a survey area. The kernel therefore has to be adapted for these particular cases to avoid strange effects at the edges and in the vicinity of "dummy readings." For this reason specialized software is often required for the processing of archaeological geophysics data as not all GIS and Remote Sensing packages provide sufficiently sophisticated tools for edges and "dummy readings."

To discuss the original concept of filters in the frequency domain, it is useful to refer to time-varying audio signals, which can be best understood as an amplitude that changes with time (figure 6.7a). However, every such audio signal also has a representation in the frequency domain (figure 6.7b), where its bass (low frequencies or slow variations) and treble (high frequencies or rapid variations) are shown on opposite ends of a spectrum.

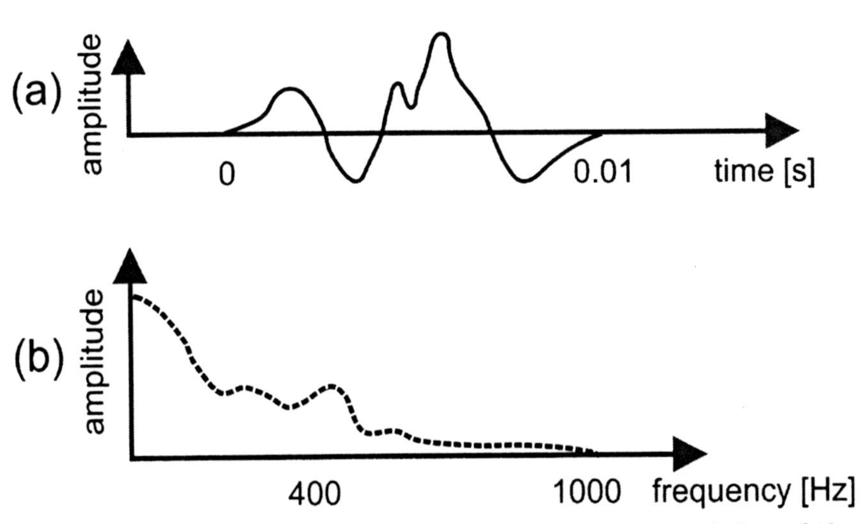

Figure 6.7. An audio signal can be visualized as a time variation of the signal's amplitude or as the strength of each frequency component. Both representations describe the same signal.

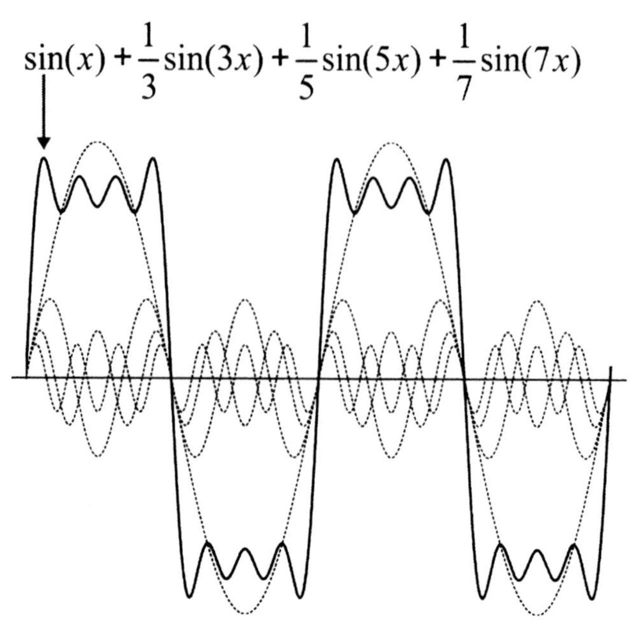

Figure 6.8. A periodic signal can be reconstructed by a combination of several sinusoidal signals with different frequency and amplitude. The example shows the approximation of a rectangular pulse.

The formal conversion from a time representation to a frequency representation is referred to as "Fourier[4] transform" and is based on the fact that any signal can be re-created by adding sinusoidal variations of various frequency and amplitude, the so-called frequency spectrum (figure 6.8).

The same procedure can be applied to signals that do not vary in time, but in space (for example, along a transect). This transformation converts from distance to a "spatial frequency" that is usually referred to as wavenumber (k) and is proportional to the inverse of the wavelength (λ): $k = 2\pi / \lambda$. A signal that has a high spatial frequency (i.e., a high wavenumber) and is varying rapidly therefore has a short wavelength. Such a spatial Fourier transform can also be applied in two dimensions, yielding as a result a two-dimensional wavenumber representation that is an equivalent expression of the original raster data and can be transformed backward and forward without data loss—for example, using the commonly implemented Fast Fourier Transform (FFT). Once the data are available in

4. Jean Baptiste Joseph Fourier, 1768–1830, French physicist and mathematician.

such a representation they can easily be filtered. If only small features are to be retained and emphasized (small wavelength), but all background variations are to be removed (long wavelength), then the high wavenumbers have to pass through the filter while low wavenumbers have to be deleted; hence a "high-pass filter" is used. By contrast, if only smooth and broad variations are to be retained (passing low wavenumbers) while small anomalies are to be removed, then a "low-pass filter" is required.

6.3.1 High-pass filtering

As mentioned above, the aim of high-pass filtering is to retain and emphasize small features while removing background variations. Broadly changing background values are typical for earth resistance data, often due to gradual variations of moisture content across a site and sometimes due to the influence of large, underlying geological structures. The archaeologically relevant anomalies, by contrast, may only show as small variations over such a background (figure 6.9a). When the whole data range is stretched to the available display values (e.g., white to black), such small changes may not be visible. By attempting to remove the underlying trend (figure 6.9b), the weak and short anomalies will be far better discernible. This is the aim of a high-pass filter, but unfortunately such an ideal implementation is not possible and instead it usually introduces new artifacts so that the actual results are more similar to those in figure 6.9c: false negative values are introduced to the sides of the original positive anomalies. This is an intrinsic problem of the computational implementation of the filter. This can best be explained with a one-dimensional example. The kernel's filter coefficients are constructed such that they estimate the background on both sides of an anomaly and subtract this estimate from the central value. In a simple example with a kernel size of five the filter coefficients would therefore be -1/4, -1/4, +1, -1/4, -1/4, which works well when applied only to the background or the central peak (figure 6.9d (1) and (2)), but when the filter kernel is starting to cover the peak, the background's estimate is wrongly influenced by the peak itself (figure 6.9d (3)), and as a result undesired filter artifacts are created. An additional effect is that the high-pass filter makes the central anomaly appear narrower and hence can create the impression of smaller archaeological features. It is essential to know about such possible artifacts when evaluating filtered data so that misinterpretations are avoided. Once a high-pass filter has

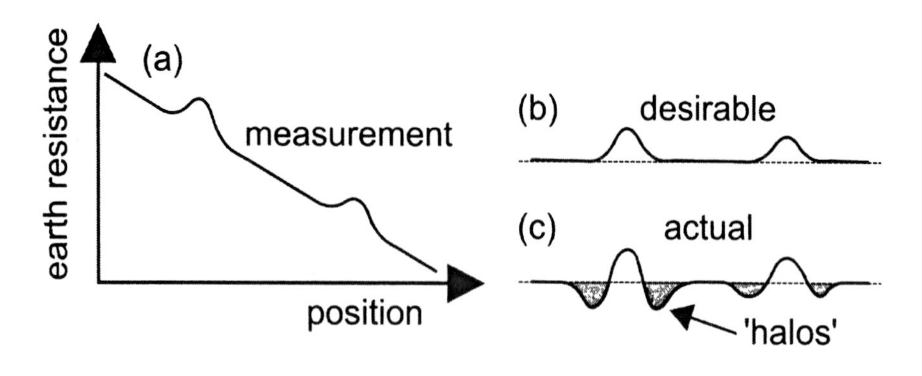

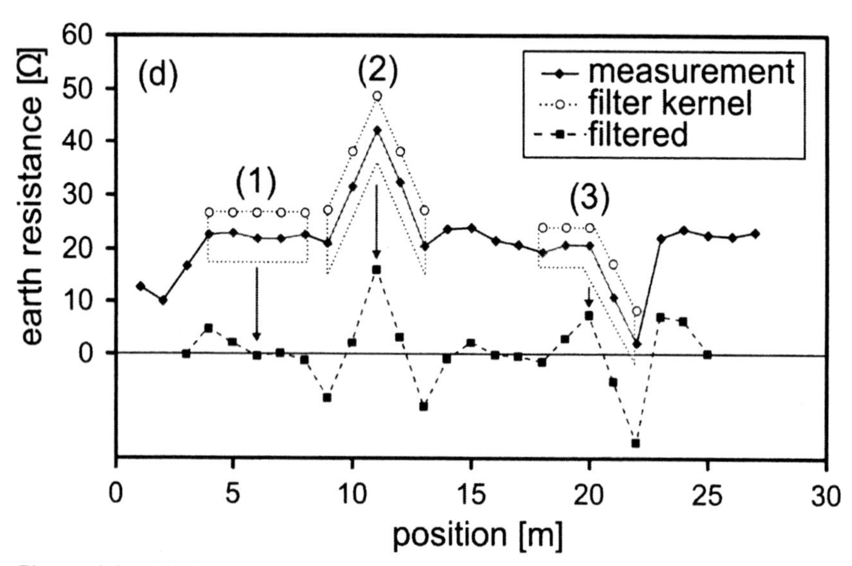

Figure 6.9. High-pass filtering of an anomaly removes the background but can introduce unwanted side effects in the form of "halos" to both sides of the anomaly: (a) measurement data with an underlying gradual variation; (b) desired result of the filter, removing the background; (c) results of the actual filter, introducing artifacts to the sides; (d) the filter kernel works well directly above the background (1) or a peak (2), but the background is estimated wrongly as the kernel moves over an anomaly (3).

revealed small and weak anomalies, it is usually possible to also discern them, albeit faintly, in the original data, possibly by adjusting the display range and contrast (see the example in section 7.4). A final interpretation diagram should then be drawn based on these re-examined original data, as the anomaly shape will be more representative.

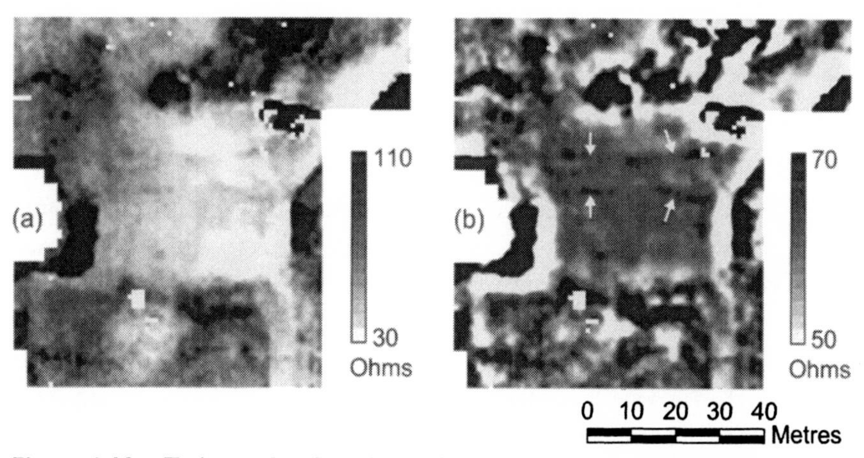

Figure 6.10. Twin-probe data from the courtyard of the Buddhist vihara at Paharpur: (a) interpolated measurements; (b) after high-pass filtering, showing two bent linear anomalies.

Figure 6.10a shows earth resistance data from the courtyard area of the Buddhist vihara at Paharpur (Bangladesh). The data are characterized by a large range of earth resistance values across the survey area, ranging from low in the middle wet areas (30 Ω, white) to high over the buried stone foundation of the original staircase (left, 110 Ω, black). After high-pass filtering two parallel linear anomalies with a kink become visible (figure 6.10b, arrows). They were interpreted as buried wall foundations for an earlier structure that formed a different alignment to the current layout. This filtering introduced very pronounced low-resistance "halos" around the areas of isolated high earth resistance, for example, to the right of the staircase's original stone foundation. In addition, the extent of the positive anomalies was changed as part of the filtering. The final interpretation diagrams were therefore derived from the original data after adjusting their display in such a way that the newly "discovered" anomalies became visible (faint traces can already be seen in figure 6.10a). However, these adjusted data displays were not suitable for presentation to the client and only appeared in the appendix of the final report. The high-pass filtered data featured in the report's main section and helped to convey the overall interpretation of results.

Most other filter functions also introduce artifacts into the resulting data. Treating the various filters that modern software packages provide as black boxes that can be applied at will can lead to serious misinterpretations of the data (Schmidt 2003). Processing data arbitrarily until the results "look good" is pointless.

6.3.2 Low-pass filtering

While a high-pass filter is used to remove unwanted data of broad background changes, the role of low-pass filtering is to suppress variations that are smaller than the investigated archaeological anomalies. In both cases the unwanted signals are commonly referred to as noise.[5] In earth resistance surveys over very stony ground, for example, successive measurements may vary considerably around a mean value due to imperfect contact with the soil. When this is not in the form of singular excessive readings that could be treated with spike removal (section 6.2.2), a low-pass filter can create a clearer outline of the underlying anomaly. The simplest implementation of a low-pass filter is as an averaging operator where the mean of all data values under the filter kernel is calculated and used to replace the central value. In a one-dimensional example with a kernel size of five the filter coefficients would either all be +1/5 or, if the central value is not used for the averaging, it would be +1/4, +1/4, 0, +1/4, +1/4. The latter filter is similar to a simple spike removal filter. The averaging of neighboring data values does not have to be uniform and, for example, a Gaussian weighting gives closer neighbors a greater influence.

The averaging process of the low-pass filter is indiscriminate and as a result the filter has the unwanted side effect of smoothing the data, reducing the "crispness" of a plot that may have initially been achieved by using a high spatial survey resolution. For pseudosection data (i.e., a two-dimensional display of apparent resistivity values), a low-pass filter can also sometimes be useful to improve the appearance of the section.

Figure 6.11 shows the northeastern survey area of the southern Thornborough henge (North Yorkshire, United Kingdom). The data were improved through spike removal, but some noisy variations remained, and it was therefore attempted to enhance the results with a low-pass filter. Figure 6.11b shows the filtered data (radius 2 m, Gaussian weighting) and figure 6.11c the difference between the original and the filtered data (that

5. This generic definition of noise as unwanted signals must not be confused with one specific unwanted signal, the "random instrument noise," which in the Germanic languages is referred to as "Rauschen."

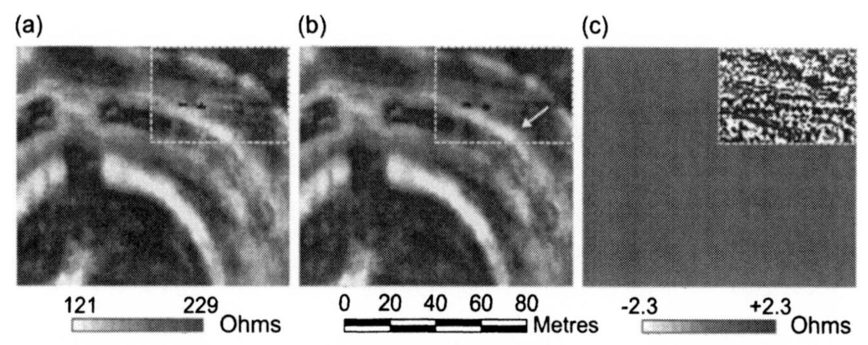

Figure 6.11. A low-pass filter was applied to the northeastern part of the twin-probe data from the southern Thornborough henge (boxed outline): (a) measurements after spike removal and block normalization; (b) results of low-pass filtering; (c) data that were removed by the low-pass filter.

is, the noisy signals that were removed). After the processing the course of the low earth resistance anomaly that forms the northern flank of the extant bank is more clearly defined (arrow), but this is at the expense of an overall loss of detail due to the filter's smoothing effect, which can be seen by inspecting the data that were removed (figure 6.11c).

6.3.3 Interpolation

Sometimes a data plot may look "noisy" simply because the variation of measurement values from one recording position to the next is too clearly discernible, creating a pixelated appearance. This is especially the case if the data were initially recorded at 1 m × 1 m sampling intervals. One way to alleviate this is to interpolate the image to a finer resolution—for example, 0.5 m × 0.5 m—so that an overall smoother appearance is achieved. The process involves the estimation of data values for positions where no measurements were taken, based on recorded readings in the surroundings. There are many different interpolation algorithms available, including bilinear, "sinc,"[6] bi-cubic, and variogram based, and most produce good results (Scollar et al. 1990; Li and Götze 1999). Some of these functions do not preserve the original data values but replace all existing data values with new and denser readings. For example, if data for a 1 m × 1 m resolution survey are collected in the center of every square cell, as

6. Defined as $\mathrm{sinc}(x) = \dfrac{\sin(\pi x)}{\pi x}$.

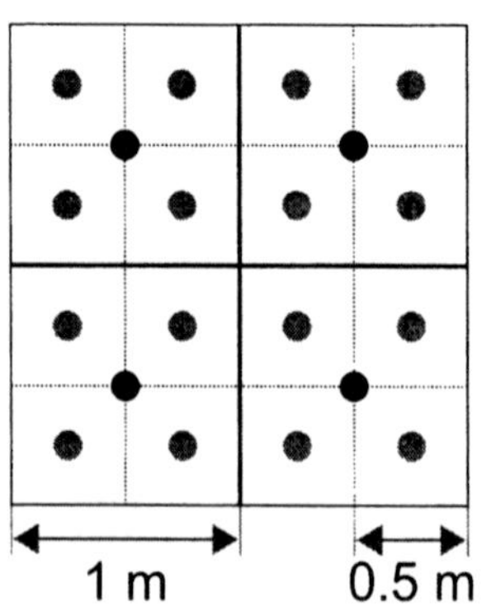

Figure 6.12. If a 1 m × 1 m survey square is interpolated to four squares of 0.5 m × 0.5 m, then the assumed measurement positions are shifted.

is usual for surveys with the Geoscan instrument, the assumed position of the new data points when interpolated to 0.5 m × 0.5 m is different, and none of them coincide with the original position (figure 6.12). Figure 6.13 shows the data from the southern Thornborough henge (North Yorkshire, United Kingdom), and the comparison between interpolated and low-pass filtered data clearly demonstrates how the former retains more sharpness while appearing less noisy than the original data.

6.4 Image processing

Data are just numbers (Schmidt and Ernenwein 2011) and have to be visualized to analyze and interpret them, most commonly as grayscale raster images. For this a "transfer function" is commonly used that determines how each data value is converted to a display color. In this process very high and very low values are usually "clipped" (i.e., mapped to the same color), and the whole process is irreversible: once the image pixels are calculated, the real measurement data cannot be reconstructed. This can create problems if one later wanted to extract subtle information from such an image with a limited number of distinct colors or shades of gray (e.g., 255). The full data precision (e.g., 120.05...455.07 Ω) cannot be re-created. When data are converted to images for display purposes, important information is therefore lost. Images are extremely useful for evaluating data, but they are

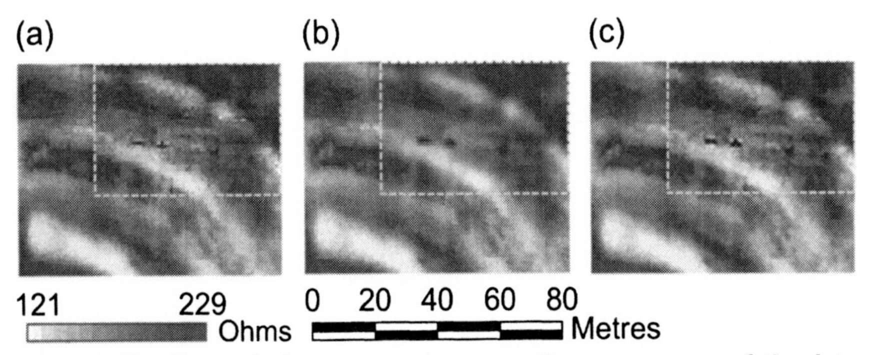

Figure 6.13. Interpolation can create a smoother appearance of the data while retaining all spatial details, demonstrated at the northeastern part of the twin-probe data from the southern Thornborough henge: (a) measurements at original resolution (1 m × 1 m); (b) low-pass filtered data at the same resolution; (c) data interpolated to 0.5 m × 0.5 m using "sinc" weighting.

no replacement for the recorded measurements. These images can only be treated with standard image processing tools that cannot take into account the geophysical nature of the underlying data. Nevertheless, this may still be appropriate to convey relevant information to users.

6.4.1 Wallis filtering

One algorithm that was found to improve images of earth resistance data is the Wallis filter. Scollar et al. (1990) showed its application to aerial photographs, but it was also found to be useful for earth resistance data. The filter uses a fairly large kernel to evaluate the range and average of all values within it and subsequently adjusts the central pixel in such a way that the brightness and contrast of all areas of the image are the same. Its implementation in Geoplot creates smoothed data that can be used as an estimate for the background variation of a dataset.

CHAPTER SEVEN
CASE STUDIES AND DATA INTERPRETATION

In this chapter several earth resistance surveys will be discussed in more detail. The emphasis will be on the data processing (see chapter 6) and the archaeological interpretation of the data. For both tasks it is important to acquire a good understanding of the parameters affecting earth resistance measurements (chapters 1 to 5) as well as an archaeological insight into the context in which the data were acquired.

7.1 Southern Thornborough henge

The survey results from the southern Thornborough henge were used throughout this book to illustrate issues related to earth resistance data and their processing. In this section, further discussion of the whole survey is provided. The investigations were started in 1992 by Arnold Aspinall and Roger Martlew and gradually were enlarged as part of our student fieldwork projects in subsequent years. Since the weather and its development prior to a survey were very different in each event, the data show considerable variations in contrast, which poses challenges when combining them.

The site covers the southernmost of the three henges at Thornborough (North Yorkshire, United Kingdom), the bank of which still stands up to 1.5 to 2 m high (Harding 2012). The three very similar henges are from the Neolithic/early Bronze Age and are classified as being of Class IIa, since they possess not only the characteristic bank and inner ditch, but also an outer ditch with two causeways leading into the monuments. The earth resistance surveys of the southern henge were undertaken to characterize it further and to test whether any internal features could be detected. In addition, fluxgate gradiometer surveys were conducted, which are not being discussed here as they added little additional information. All data were collected with an RM15 earth resistance meter using a conventional 0.5 m twin-probe configuration. The majority of the site was investigated at a spatial resolution of 1 m × 1 m over 20 m square data grids (20 m × 20 m @ 1 m × 1 m) covering an area of approximately 180 m × 220 m. In addition, a smaller area at the northwestern terminal of the inner ditch was investigated at a higher resolution (10 m × 10 m @ 0.5 m × 0.5 m) over 20 m × 50 m.

The raw data of the main area showed the different individual blocks of the survey very clearly (figure 7.1a), and due to the considerable range of values (84...334 Ω) only the very pronounced inner ditch could be discerned. Using the automated edge matching algorithm of the Contors software (Haigh 1992), the data became much more intelligible and could be displayed at the narrower range of 112...234 Ω (figure 7.1b; same data as figure 6.3a), so that some weaker anomalies also became visible. However, there remained clear differences in the range of data values for each of these blocks, since different soil moisture conditions and possibly changes in the separation of remote electrodes led to different contrasts. Each contiguous data block was therefore normalized to a range of 50...150 Ω (see section 6.2.1 and the discussion therein about the remaining mismatches of contrast). Subsequently, all data grids were individually edge matched with the automated routine from Contors to minimize remaining mismatches (figure 7.1c, same data as figure 6.3b). One might have wanted to adjust whole blocks instead of individual grids by masking each block and adding an appropriate offset value to it. This somewhat elaborate approach was attempted for one block edge, and the results were found to be similar to the automated balancing of all grids, so that the latter routine was used for the whole site. The data were then further improved in Geoplot by applying a spike removal filter that removed some of the erroneous data in the northeastern area.

The improved data showed considerable detail and were well suited for analysis and interpretation. However, for inclusion in a report for the end users and for printing (which has a far weaker dynamic range than

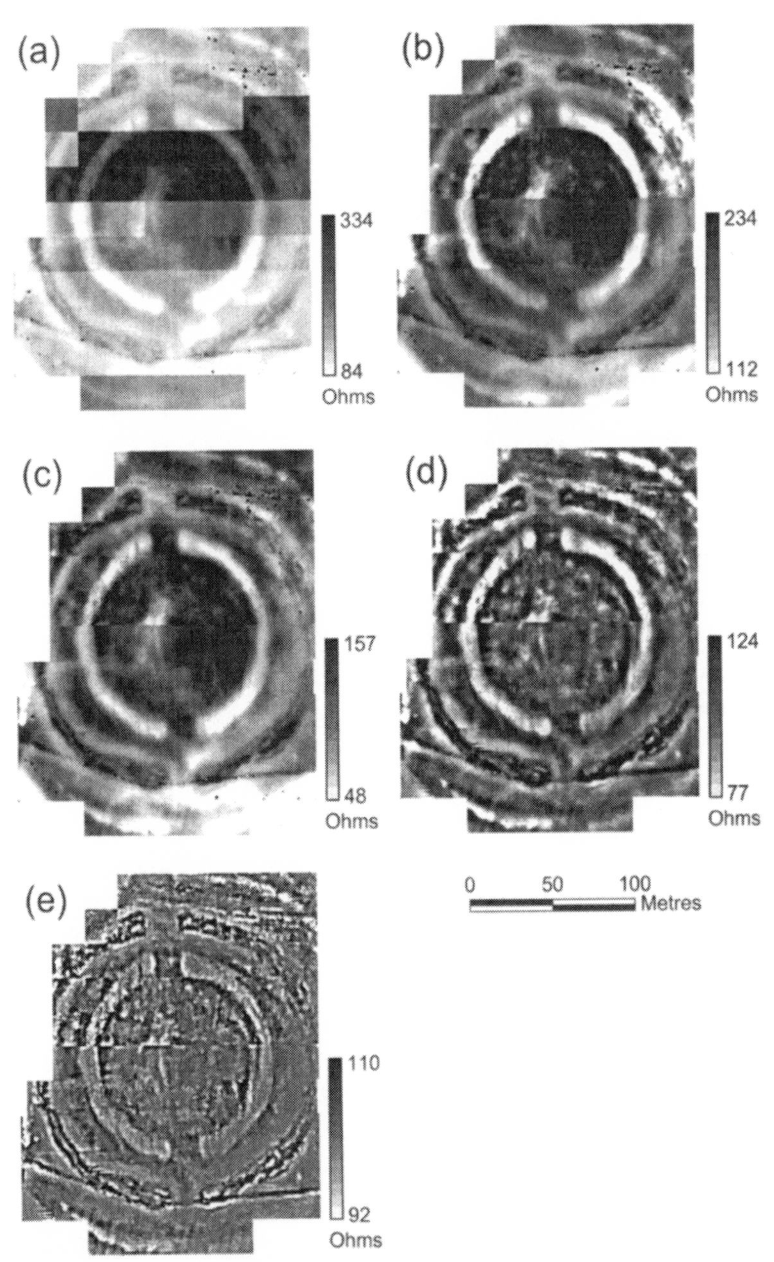

Figure 7.1. A sequence of processing steps was applied to the twin-probe data from the southern Thornborough henge: (a) raw data; (b) data grids were edge matched; (c) data blocks were normalized and then all data grids were edge matched; (d) data from (c) were high-pass filtered; (e) the Wallis filtered data of (c) were subtracted from (c).

modern computer displays), some further enhancement of these features was necessary. The data were therefore high-pass filtered (filter kernel of 21 × 21, Gaussian weighting; figure 7.1d), but since this operation enhances small variations it also highlighted the remaining differences in contrast between the blocks too strongly and was hence disregarded. Instead, a weak Wallis filter was used to estimate background variations (filter kernel of 5 × 5, uniform weighting), and this was subtracted from the enhanced data. The results (figure 7.1e) provided a clear depiction of all anomalies that were considered to be relevant for interpretation and were hence used to illustrate the overall survey results. As discussed in section 6.3.2, the low-pass filtering of the data resulted in considerable loss of detail and was not implemented, as it would only have made the overall results "look nice" at the cost of interpretable anomalies.

The data were subsequently integrated into a project GIS as georeferenced images of enhanced and processed results (figure 7.2a), combining the datasets of the two different spatial resolutions (see above) and overlying them with a hachure representation of the still visible earthworks (i.e., banks and slight depressions of the ditches). The interpretation diagrams were then drawn based on the enhanced data and are shown in figure 7.2b. Anomalies were traced as polygonal features, as all anomalies have a width and cannot be represented appropriately at intermediate scales as line drawings.[1] The two parts of the inner ditch can be seen very clearly as low earth resistance anomalies (1), having nearly square terminals at both ends. The low resistance is a result of the better moisture retention of these features compared to the background. Their extent as deduced from the earth resistance results is somewhat larger than what is estimated from the weak topographical depressions, shown in the hachure plot. The northeastern terminal of the western ditch segment shows some additional high-resistance anomalies, which was the reason for the high-resolution survey over this area. A high-resistance anomaly is cutting through the ditch at an oblique angle (2), and a subsequent excavation found this to be the remains of a stone wall, probably of more recent agricultural origin (hence the oblique angle). At close inspection it could also be traced further south in the lower resolution data. A second and weaker high resistance anomaly (3) can be seen 6 m west of it. Just east of this ditch-terminal an

---

1. When displayed as an overview diagram, however, a line drawing may be useful.

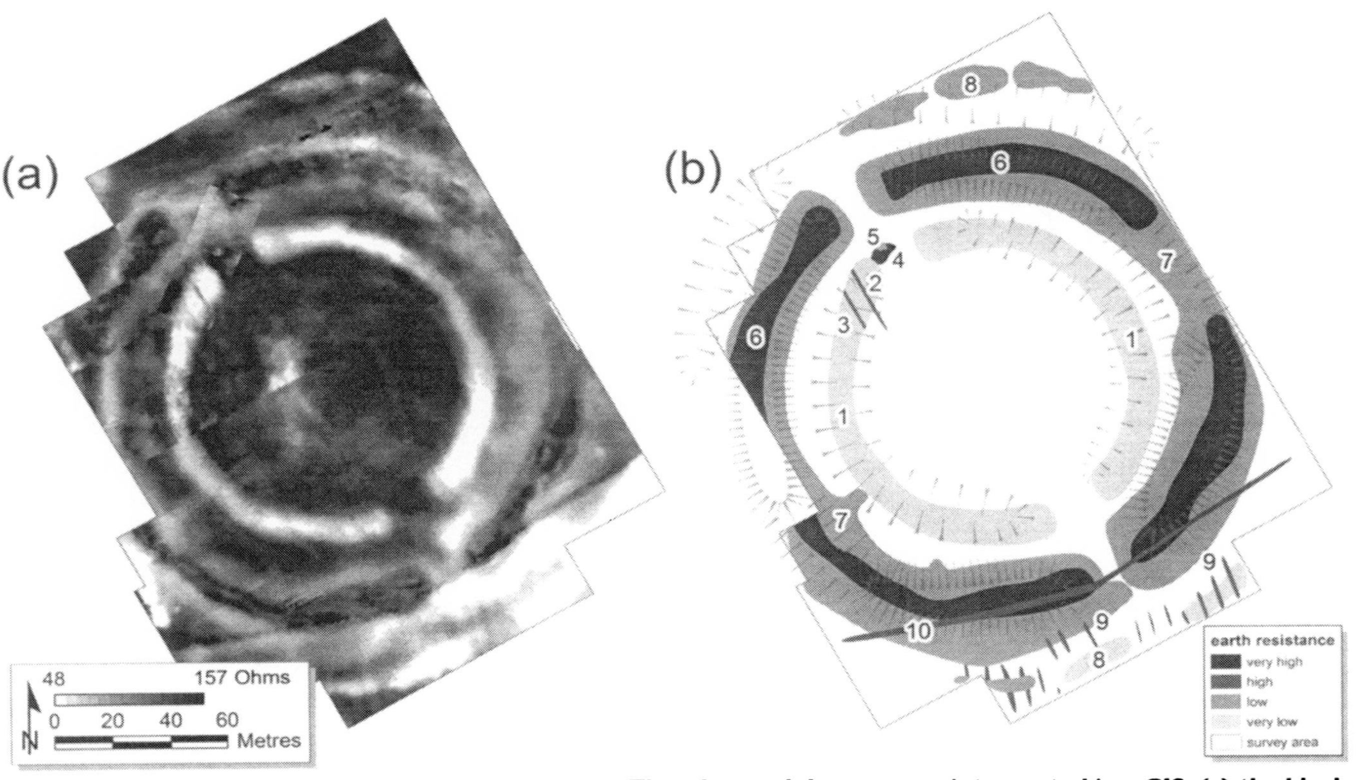

Figure 7.2. The twin-probe data from the southern Thornborough henge were interpreted in a GIS: (a) the block normalized data were georeferenced; (b) interpretation polygons of the anomalies with hachures representing the earthworks.

extended area of high resistance is visible (4), being cut by a round, low-resistance anomaly of 2 m diameter (5). It is tempting to interpret these as stone packing (4) around a large posthole (5), but there is no equivalent anomaly visible on the opposite ditch terminal, so the final interpretation of these anomalies will have to await further excavations.

The extant bank of the henge is characterized by very high resistance on its top (6) with low resistance all around its flanks (7) and very low values in the inner ditch (see section 5.3.3). This may be a purely topographical effect of water draining from the top, down the slope, and accumulating in the ditch, but the very clear and confined outline of the high-resistance anomaly (6) that shows best in the Wallis filtered data (see figure 7.1e) could also indicate a rubble core in the bank of the henge. The outer ditch of the henge is also visible as a low-resistance anomaly (8), where the survey extended far enough. However, it appears far more segmented than the inner ditch, and this may reflect the method of its construction. In the far south of the dataset several parallel high-resistance anomalies mostly cross the outer ditch (9) and are probably due to plowing after the ditch had already filled up. A linear kinked anomaly south of the main bank is the line of a fence (10).

7.2 Newstead Roman fort

The Roman fort at Newstead and its hinterland was investigated by geophysical means from 1987 to 1998 by Rick Jones and Paul Cheetham (Clarke and Wise 1999). Pseudosection data from the project were already discussed in section 4.4, and this section focuses on the earth resistance area surveys of the eastern part of the main fort.

In September 1992 a total of 175 data grids of 20 m square size were collected (20 m × 20 m @ 1 m × 1 m) with an RM15 earth resistance meter using a standard 0.5 m twin-probe array, covering part of a 260 m × 260 m field. There were only small edge mismatches between the individual survey blocks (figure 7.3a), which were eliminated with the automated edge-matching algorithm of Contors (Haigh 1992). However, as Contors was unable to load all data grids simultaneously, they were grouped into blocks, edge matched separately, and then the blocks were stitched together. The data were subsequently imported into Geoplot, and a standard spike removal was applied, producing the improved dataset

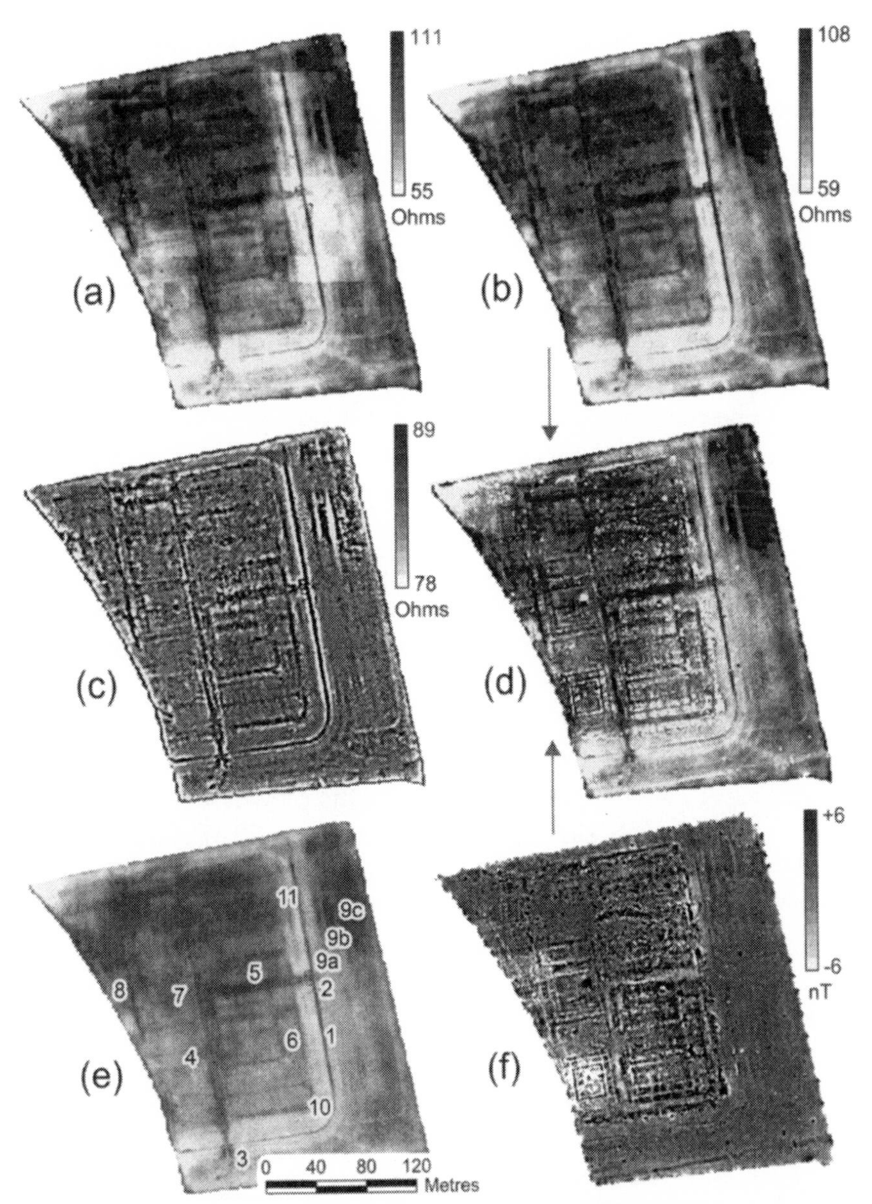

Figure 7.3. A sequence of processing steps was applied to the twin-probe data from Newstead Roman fort: (a) raw data; (b) data grids were edge matched and spikes removed; (c) Wallis filtered data were subtracted from (b); (d) combination of scaled earth resistance data (b) and magnetometer data (f); (e) feature labels as referred to in the text; (f) fluxgate gradiometer data of the same area.

shown in figure 7.3b. Although all relevant anomalies were visible in this dataset they were further enhanced for presentation purposes through background removal. It was found that the best-looking results were achieved by subtracting a smoothed version of the data, obtained with a Wallis filter of 7 × 7 kernel size (uniform). The results were then interpolated to 0.5 m × 0.5 m (figure 7.3c).

The data show many of the features highlighted by Richmond in his analysis of the Antonine fort (Richmond 1949–1950: figure 5). The foundations of the stone wall surrounding the fort are clearly visible as high-resistance anomalies (figure 7.3e (1)), including the outline of the towers of the eastern gate, the *porta praetoria* (2). The southern gate (3) seems to be constructed differently, as the main high-resistance anomaly lies outside the wall, but its shape cannot be delineated accurately; maybe a higher spatial survey resolution would have been required. Inside the fort the *via principalis* (4), running north-south, and the *via praetoria* (5), extending from the eastern gate up to the *via principalis*, show as broad high-resistance anomalies, probably caused by stone pavements. The anomalies of these roads do not continue outside the fort, indicating a different method of construction there.

North and south of the *via praetoria* are the outlines of the soldiers' barracks (6), while on the western side of the *via principalis* the 40 m × 40 m outlines of the *principia* can be seen, having six cells at the back (west) and five in the north, surrounding a central courtyard (7). This structure would have incorporated the shrine of the legionary standards of the *Legio XX*, stationed there. A linear high-resistance anomaly is visible to the west of the *principia* (8) and can be identified as the dividing wall that was constructed to split the fort in the second Antonine phase (Richmond 1949–1950: 19).

Outside the fort three parallel low-resistance anomalies are visible in the east and south, forming the outer defensive ditches (9a–c). Inside the eastern and southern section of the main wall a broad band of low resistance clearly separates the fort's interior from the surrounding wall (10). This is interpreted as the footing for the internal rampart that supported the wall; the thin, high-resistance anomaly in its middle (11) may be due to stones laid during its construction. Notably absent are any anomalies related to the commandant's house (*praetorium*) expected south of the *principia*. Such anomalies are instead clearly visible in the magnetometer data (figure 7.3f), where they show as negative outlines of the *praetorium*'s walls, with positive magnetic fill in the individual rooms. This is interpreted as the result of strong burning in the end phase of the fort so that highly magnetized material, including tiles from a collapsed roof, accumulated in the rooms, thereby forming a positive contrast against the nonmagnetic stones of the wall. One possible reason for the lack of earth resistance anomalies over this *praetorium* is an earlier excavation of this structure that may have removed or disturbed the top courses of stones but left enough of the magnetized material to create magnetic anomalies. The southeastern barrack (6) is also characterized by negative magnetic anomalies, while the *principia* has negative anomalies in the north and positive anomalies in the south, indicating a reversal of the magnetic properties, possibly as a result of different building materials or different burning events. The fort's stone wall creates only faint magnetic anomalies, while extended bipolar anomalies are visible around the inner edge of the internal rampart, probably caused by ovens or furnaces.

It is desirable to show all geophysical anomalies in one data display, and for this purpose the two datasets were combined (figure 7.3d). While the main structural earth resistance anomalies are positive (for example, the cells of the *principia*), the *praetorium* has negative magnetic anomalies and therefore the magnetic data were inverted before all data were combined. Through addition and multiplication, they were then stretched to the same numerical range as the earth resistance data (i.e., from -6...+5 nT to 65...97 nT), and, finally, both datasets were averaged. The results provide a comprehensive overview of all geophysical data and their spatial location. However, it is no longer possible to determine with which geophysical method a particular anomaly was recorded, and more elaborate visualization methods are necessary for a joint interpretation of the data. A simple display side-by-side, as in figure 7.3, does not allow the accurate investigation of their spatial relationship. By using colors and distinct visual classes (Schmidt 2001), it becomes possible to display two or more datasets simultaneously while retaining full spatial information for each.

For example, by designing a two-dimensional color scheme earth resistance and magnetometer data can be displayed in one image and distinguished by an interpreter based on the respective color of each pixel. This is similar to the false-color display of multiband satellite data that was found to be also of use for archaeological geophysics data (Kvamme

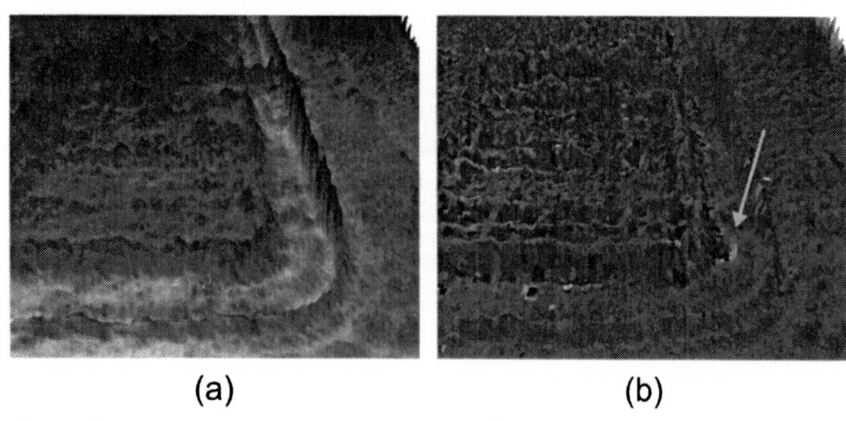

Figure 7.4. Twin-probe and fluxgate gradiometer data from the southeastern corner of Newstead Roman fort are displayed simultaneously using two different visual classes: (a) a three-dimensional visualization of the earth resistance data; (b) a grayscale image of the magnetometer data is draped over the three-dimensional earth resistance data, which allows one to distinguish the data sources and shows that some of the bipolar magnetic anomalies are located directly over the inner rampart (arrow).

2006). An alternative approach is to use a combination of grayscale plots, contours, and three-dimensional displays to distinguish different data sources, as these can be overlaid while still retaining their distinctiveness. Figure 7.4 shows an example from the southeastern corner of the fort's main wall and its internal rampart. The earth resistance data were used to generate a three-dimensional display (figure 7.4a), where the positive resistance anomaly of the wall forms a clear linear structure, while the negative anomaly of the rampart shows as a trough bounded on the west by the fort's interior. Once a grayscale plot of the magnetometer data was draped over this display (figure 7.4b), it became clear that some of the extended bipolar magnetic anomalies were actually located inside the rampart and were hence interpreted as ovens or furnaces built into the rampart, which is an unexpected finding. Further analysis of the combined data for this site can be found elsewhere (Schmidt 2001).

7.3 Warter priory

The Augustinian priory of Warter (East Yorkshire, United Kingdom) only survives as low earthworks in a pasture field. While these slight rises provide a fair impression of the priory and its associated farm buildings, for example, from aerial photographs (Stoertz 1997), geophysical surveys were undertaken by Duncan Brown in 1998 to gain further insights. The area of the priory church was of particular interest as the village's parish church had been built partly on top of it and it was unclear how much of the original church had survived in the ground. Fluxgate gradiometer and earth resistance surveys were undertaken to the east of the current church with the additional aim to investigate the impact of spatial survey resolution on the interpretability of the geophysical results. Since magnetometers passively measure already existing anomalies of the earth's magnetic field (Aspinall et al. 2008), it was no surprise that every increase in spatial sampling resolution led to sharper data that showed ever more detail. The interpretation diagram resulting from the magnetometer data of the priory church is shown as an outline in figure 7.5a.

An RM15 earth resistance meter was used with a standard 0.5 m twin-probe array to survey twenty-six data grids using three different spatial resolutions: 1 m × 1 m, 0.5 m × 0.5 m, and 0.25 m × 0.25 m. The area covered decreased with increasing resolutions due to the additionally required survey efforts (the respective survey areas are outlined with thin, broken lines in figure 7.5). The data did not require edge matching, and the only necessary data improvement was spike removal. Further processing provided no additional insights and was therefore omitted.

The 1 m resolution data (figure 7.5a) outline the main features of the church and show a rectilinear high-resistance anomaly that delineates the platform on which it is positioned. This narrow anomaly (ca. 1.5 m wide) separates an area of lower resistance around the church (ca. 13 Ω) from the higher resistance values in the lower-lying eastern area (over 30 Ω) and is interpreted as a retaining wall that also helped to hold moisture in the platform. By improving the resolution to 0.5 m the spatial definition of the church's anomalies improved considerably (figure 7.5b). At this resolution the layout of the church can be seen nearly as clearly as in the magnetometer data. Buttresses are visible to the north and the east, while a small annex, probably the sacristy, clearly shows in the south. The eastern wall produces a particularly high-resistance anomaly with a slightly amorphous shape, possibly resulting from a backfilled excavation trench that had been hinted at in documentary sources. Just west of the eastern wall is a smaller high-resistance anomaly that is interpreted as the base of the altar.

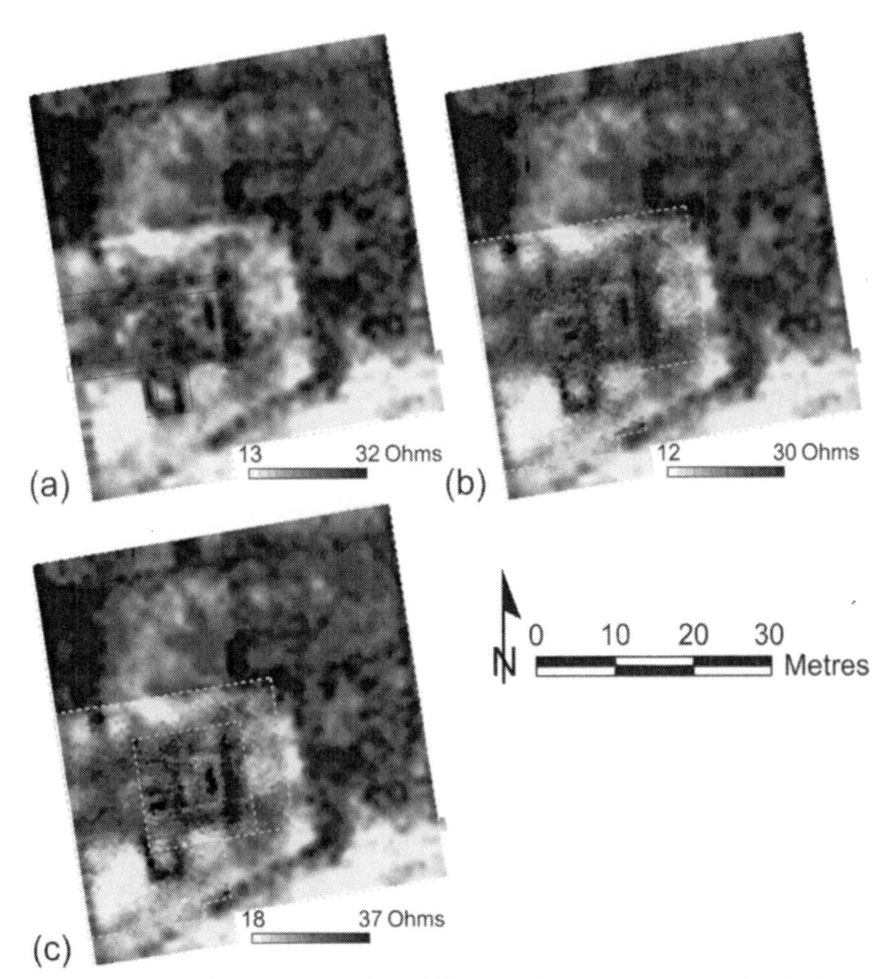

Figure 7.5. Twin-probe data from Warter priory were acquired with three different spatial resolutions. The outline of the respective survey areas is indicated with broken lines: (a) 1 m × 1 m, with an interpretation outline derived from the fluxgate gradiometer data; (b) 0.5 m × 0.5 m; (c) 0.25 m × 0.25 m.

Aligned with the eastern wall of the sacristy annex in the south is the foundation of a dividing wall that crossed the nave. Perpendicular to it are two small, linear anomalies aligned with the long axis of the nave. These are most clearly defined in the data with the highest spatial resolution (figure 7.5c), but their interpretation remains unresolved. These latter high-resolution data add only a little more definition to the shape of the anomalies, since all surveys were undertaken with a 0.5 m twin-probe

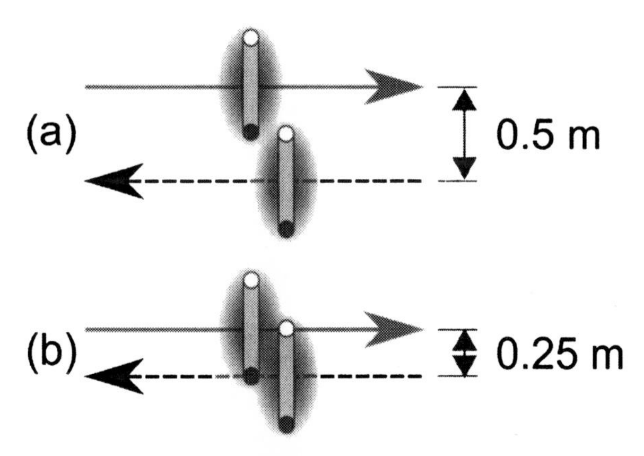

Figure 7.6. If traverses are closer together than the electrode separation, the resulting measurements probe partially the same ground. This is shown for a twin-probe array with 0.5 m separation of the mobile electrodes: (a) a 0.5 m separation of transects leads to little overlap of the measured ground; (b) a 0.25 m separation of transects leads to considerable overlap of the measured ground.

array. At that high spatial resolution measurements on adjacent survey transects probed partially the same ground and did not add much new information (figure 7.6). The electrode separation of the array would have had to be reduced to the transect spacing (in this case 0.25 m) to obtain sharper data, but this would have led to a shallower investigation depth and hence to incompatible data for the comparison of different sampling densities (Parkyn 2010).

These results clearly highlight the importance of choosing a spatial resolution for the survey that is in accordance with the archaeological objectives of the investigation. If only the location and broad outline of a structure with strong resistivity contrast is of interest, a low sampling resolution may be sufficient. For a more detailed analysis of the shape of buried features the resolution has to be increased, but not beyond the size of the electrode array used.

7.4 Tilaurakot

Data from Tilaurakot (Terai, Nepal) were briefly presented in section 6.2.2 to demonstrate the effect of spike removal. In addition, several other processing steps were needed to fully analyze these data, as will be explained in this section. The site of Tilaurakot consists of a fortified low citadel mound of approximately 500 m × 400 m size and is surrounded by a series of associated monuments (Schmidt et al. 2011). In 1899 the ruins at Tilaurakot were investigated, surveyed, and excavated by P. C. Mukherji of the Archaeological Survey of India, and he concluded that the remains represented Kapilavastu, the capital of King Suddhodana, Buddha's father, and as such would have been the site of the palace in which Siddhartha Gautama grew up as a prince before renouncing all worldly goods. Although all circumstantial evidence seems to support this attribution, a firm proof is impossible as there will be no archaeological evidence that could undoubtedly be assigned to Siddhartha Gautama. The geophysical investigation instead focused on the remains of the latest phase of this heavily overgrown site, which is unique and important in the wider region. Several areas were cleared for fluxgate magnetometer and earth resistance surveys, and areas 3 and 7, just inside the western gate, are those being reported here (Schmidt et al. 2011).

The contiguous area was covered with fifty-five data grids of 10 m square size using an RM15 earth resistance meter with a 0.5 m twin-probe array (10 m × 10 m @ 0.5 m × 0.5 m). The data did not require edge matching, and a conventional spike removal filter was applied to create the improved data (see section 6.2.2, figure 6.4). Since many mature trees and other obstacles made the survey very cumbersome, the data plot had a peculiar shape and many small data holes. These were marked as "dummy readings" during the survey, for example, where trees were located (figure 7.7a). To improve the appearance of the data a stitching algorithm was used based on ideas by Irwin Scollar (Scollar et al. 1990). For this procedure a low-pass filter with a kernel size of 3 × 3 was repeatedly moved over the dataset to estimate smooth replacement values for missing readings, until all data holes were filled. After stitching the data holes in this way, the spike removal filter was applied a second time, improving the data further (figure 7.7b). The outlines of some building foundations were visible in the northern part of the data, but since the visualization used a wide data range to display results from the northern as well as the southern section, details are hard to distinguish (overall range 6...20 Ω). Displays with different ranges were therefore created: 8...30 Ω to highlight features in the northern part (figure 7.7c) and 6...12 Ω for the southern part (figure 7.7d).

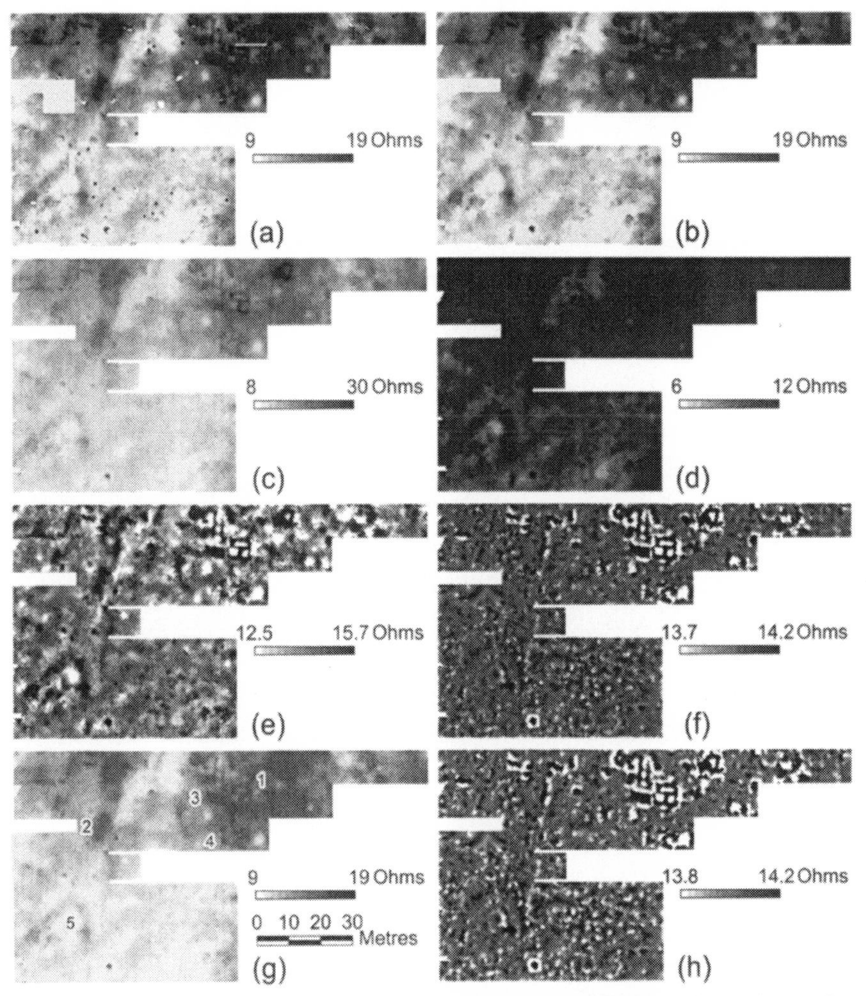

Figure 7.7. A sequence of processing steps was applied to the twin-probe data from the citadel at Tilaurakot: (a) raw data; (b) stitched and spike removal; (c, d) data from (b) with different display ranges; (e) the data from (b) were high-pass filtered by subtracting low-pass filtered data; (f) an alternative high-pass filter was applied by subtracting Wallis filtered data from (b); (g) feature labels are overlaid on the data from (b); (h) low-pass filtering of (f) removes some noise.

Several distinct anomalies, mainly in the north, were visible in these images, but the results would be more easily conveyed if all information were visible in one display. It was therefore attempted to remove the background variation with a filter. This would usually be done with a high-pass filter (section 6.3.1), but the background variation on this site proved to be so broad that the Geoplot high-pass filter was insufficient. As an alternative high-pass filter, the background was estimated with a large low-pass filter (kernel size of 21 × 21, Gaussian) and then subtracted from the improved data (figure 7.7e). Different results were achieved by subtracting Wallis filtered data (kernel size of 5 × 5, uniform) from the improved data (figure 7.7f).

Both of these high-pass filters highlighted the rectilinear cells in the north (figure 7.7g (1)), especially the Wallis filter. However, the latter also removed a rectangular area of high resistance, leaving only a faint positive outline in its place (2). Both high-pass filters created pronounced low-resistance halos around the main high-resistance anomalies. It was decided that for a final presentation of the data the noise between the strong rectangular anomalies should be reduced, and a low-pass filter was applied to the Wallis filtered data. However, the standard low-pass filter parameters of a kernel size of 5 × 5 (i.e., 2.5 m × 2.5 m) with a Gaussian weighting smoothed the corners of these structures too much, making them look more round than rectangular. To use a smaller filter size the data were first interpolated to a resolution of 0.25 m × 0.25 m so that the Gaussian low-pass filter with kernel size 5 × 5 only covered an area of 1.25 m × 1.25 m. The resulting data are shown in figure 7.7h, having less noise while preserving the rectangular shape of the northern anomalies. Overall, the filters were selected in such a way that the processed data looked similar to what had already been detected using displays with different data ranges (see above).

Several rectangular structures in the northern part of the survey area (1) are interpreted as substantial brick foundations, very similar to the outlines of so-called palace structures excavated at Mound VII in the 1970s. In addition to the cell-like structures there is also a rectangular anomaly (3), which is most likely a platform. The Wallis filtered data create the impression that a cluster of several broken anomalies (4) is also part of this layout. All these anomalies are on a very similar alignment to each other, with an overall bearing that is a few degrees further east than the axis of the western gate, which lies just west of the survey area. By contrast, the direction of anomalies in a broad corridor in the western part of the survey block, including the platformlike anomaly (2), is quite different, trending southwest to northeast. One might even interpret the southwestern part of this area as showing a Y-shaped split of possible paths at a large, round, low-resistance anomaly (5). It is possible that this area formed a thoroughfare from the western gate to the northern and eastern parts of the site.

There are also many round, high-resistance anomalies, and it is likely that these are pits related to the modern vegetation of the site. The fluxgate gradiometer data for this survey area are far less clear and did not allow drawing conclusions about the layout and internal organization of the citadel. It was therefore crucial that the high-resolution earth resistance data were collected, and advanced high-pass filtering of the data helped to visualize the results.

CHAPTER EIGHT
CONCLUSION

Earth resistance techniques are not new; they have been used in archaeology for at least two decades, mainly in the form of twin-probe area surveys. This was mainly due to the ease of use and availability of reliable equipment, such as Geoscan's earth resistance meters RM4, RM15, and RM85, and the TR/CIA resistivity meter (also an earth resistance meter) from the Council for Independent Archaeology. The effectiveness of the RM15 was enhanced considerably with the introduction of the MPX15 multiplexer, which allows measurements with different electrode configurations at each insertion point. Still, overall survey speed remained fairly slow due to the need for manual insertion of electrodes into the ground.

The advent of wheeled devices made investigations much faster, starting with the French RATEAU square array system and its variants, including Leckebusch's tractor (Leckebusch 1998), and later the Geoscan MSP40 platform. The combination of a towed multidepth device (ARP from Geocarta; see section 2.4.3) and GPS recording of measurement positions opened up new and exciting avenues for mapping large areas with earth resistance measurements, similar to motorized magnetometer and GPR systems.

This development may help to reverse a trend seen in recent years in the field of commercial geophysical surveys—namely, to favor magnetometer surveys over everything else. In many countries geophysical surveys are used to assess buried archaeological heritage in advance of building projects and thereby help with the preservation of important monuments. However, a strong focus on low costs means that the most appropriate techniques are not always used. In practice, magnetometer surveys were often favored even if other techniques, such as earth resistance investigations, would have been more appropriate. Fast-wheeled devices may offer new opportunities given the time efficiencies that can be gained. They were, for example, used in advance of a very large motorway-widening scheme in Italy, alongside conventional magnetometer surveys (Campana and Dabas 2011).

The success and prevalence of magnetometer surveys have also led some archaeologists and planning officers to equate "geophysics" with "magnetometer surveys." This book showed that earth resistance methods can be highly valuable, and it also explained where these may be superior to other methods. How does one decide between different geophysical methods in archaeology? As in many technical fields, which require knowledge, skill, judgment, and practice, there are no simple rules as to when to use what technique ("if A, then B"). As with all geophysical methods, there are so many different parameters to consider that generally applicable rules are virtually impossible to design. In addition, most of the time archaeological prospecting is used because it is not yet known what exactly lies in the ground. Hence, if a survey methodology is drawn up too narrowly for finding one type of archaeological remains (e.g., Roman stone structures), it is possible that the survey may miss many other features that would also be of great interest but that are not yet known to exist in the area (e.g., prehistoric ditched enclosures).

This book provides background information, illustrated with case studies, to help the user acquire the knowledge and skills necessary to become a competent archaeological geophysicist who with time will develop the practice and judgment to pick the right method for investigating a site. The background information and case studies combined can be used to make informed choices when assessing the suitability of earth resistance surveys for archaeological evaluations. Such a considerate approach should be adopted whenever possible. A table that lists archaeological features against possible geophysical techniques can be found elsewhere (David et al. 2008), and Somers et al. (2003) developed a software-decision tool, mainly for the United States.

As mentioned in the introduction, this book does not discuss what to do with one's earth resistance data after they are processed and interpreted, since archiving and documentation are well covered elsewhere

(Schmidt and Ernenwein 2011). Reusing data, either one's own or others', requires good documentation about all the relevant survey parameters. As was shown in chapter 1, having records of weather and survey conditions is crucial for the processing and interpretation of archived measurements. Exchanging archived data of successful and unsuccessful earth resistance surveys with colleagues will help to compile regional studies that evaluate the conditions under which certain parameters work best (Jordan 2009). The lack of geophysical anomalies cannot be taken as a prediction for the absence of archaeological features (the "dilemma of negative evidence"), but a large corpus of survey results allows a far better understanding of these issues and provides ways of making probability-guided estimates.

While earth resistance area surveys were used regularly and consistently in the last decade and are probably about to make great advances in survey speed and area coverage, the changes we see to the use of resistivity imaging (ERI) in archaeology are even more dramatic. Resistivity imaging was only used on a few archaeological projects, mainly for research purposes, and often just with one or two imaging sections. The time and cost for collecting and processing the data did not seem to justify the archaeological insights that could be derived from the results. It was hoped that GPR surveys would allow visualization of the three-dimensional subsurface in detail, and in many cases this has been achieved. However, the depth penetration of GPR signals has been far less than what was hoped for, especially in clay soils. Great advances in instrumentation and data processing have allowed for a very promising comeback of electrical imaging techniques. Using fast multichannel earth resistance meters makes it possible to collect data for three-dimensional resistivity tomography and to process them in a reasonable time, even for features with difficult shapes and unconventional electrode arrangements. The availability of equipment and software will further improve, and it is very likely that electrical methods will make considerable contributions to the three-dimensional visualization of archaeological remains.

The prospects for future developments are exciting, and although earth resistance techniques previously seemed to be overshadowed by magnetometer and GPR surveys, they have quietly matured and are now a key tool for archaeological geophysicists.

BIBLIOGRAPHY

Al Chalabi, M. M., & A. I. Rees
1962 An experiment on the effect of rainfall on electrical resistivity anomalies in the near surface. *Bonner Jahrbücher* 162: 266–271.

Apparao, A., T. G. Rao, R. S. Sastry, & V. S. Sarma
1992 Depth of detection of buried conductive targets with different electrode arrays in resistivity prospecting. *Geophysical Prospecting* 40(7): 749–760.

Apparao, A., & G. S. Srinivas
1995 On: "Depth of investigation of collinear electrode arrays over homogeneous anisotropic half-space in direct current methods," by B. B. Bhattacharya and M. K. Sen (May 1981 *Geophysics* 46: pp. 768–780). *Geophysics* 60(6): 1936–1941.

Aspinall, A., C. F. Gaffney, & A. Schmidt
2008 *Magnetometry for archaeologists (Geophysical methods for archaeology).* Lanham, MD: AltaMira Press.

Aspinall, A., & J. Lynam
1968 Induced polarization as a technique for archaeological surveying. *Prospezioni Archeologiche* 3: 91–93.
1970 An induced polarization instrument for the detection of near-surface features. *Prospezioni Archeologiche* 5: 67–75.

Aspinall, A., & M. K. Saunders
2005 Experiments with the square array. *Archaeological Prospection* 12(2): 115–129.

Barker, R.

1992 A simple algorithm for electrical imaging of the subsurface. *First Break* 10(2): 53–62.

Barker, R. D.

1989 Depth of investigation of collinear symmetrical four-electrode arrays. *Geophysics* 54(8): 1031–1037.

Bevan, B. W.

2000 *The pole-pole resistivity array compared to the twin electrode array*. Geosight Technical Report No. 6.

Burger, H. R., & D. C. Burger

1992 *Exploration geophysics of the shallow subsurface*. Englewood Cliffs, NJ: Prentice Hall.

Campana, S., & M. Dabas

2011 Archaeological impact assessment: The BREBEMI project (Italy). *Archaeological Prospection* 18(2): 139–148.

Christensen, N. B., & K. I. Sørensen

1998 Surface and borehole electric and electromagnetic methods for hydrogeological investigations. *European Journal of Environmental and Engineering Geophysics* 3(1): 75–90.

Clark, A. J.

1968 A square array for resistivity surveying. *Prospezioni Archeologiche* 3: 111–114.

1980 *Archaeological detection by resistivity*. Thesis. University of Southampton.

1996 *Seeing beneath the soil: Prospecting methods in archaeology* (Revised paperback edition). London: Batsford.

Clarke, S., & A. Wise

1999 Evidence for extramural settlement north of the Roman fort at Newstead (Trimontium), Roxburghshire. *Proceedings of the Society of Antiquarians in Scotland* 129: 373–391.

Conyers, L. B.

2004 *Ground-penetrating radar for archaeology (Geophysical methods for archaeology 1)*. Walnut Creek, CA/Oxford: AltaMira Press.

Cook, H. F., & D. L. Dent

1990 Modelling soil water supply to crops. *CATENA* 17(1): 25–39.

Corwin, D. L., & S. M. Lesch
2003 Application of soil electrical conductivity to precision agriculture: Theory, principles, and guidelines. *Agronomy Journal* 95(3): 455–471.

Cott, P. J.
1997 *The effect of weather on resistivity measurements over a known archaeological feature.* MPhil Thesis. Department of Archaeological Sciences. Bradford: University of Bradford.

Dabas, M., T. Jubeau, D. Rouiller, J.-M. Larcher, S. Charriere, & T. Constant
2012 Using high-resolution electrical resistivity maps in a watershed vulnerability study. *First Break* 30(8): 51–56.

David, A., N. Linford, & P. Linford
2008 *Geophysical survey in archaeological field evaluation* (2nd edition). Swindon: English Heritage.

Donahue, R. E., & W. A. Lovis
2006 Regional settlement systems in Mesolithic northern England: Scalar issues in mobility and territoriality. *Journal of Anthropological Archaeology* 25(2): 248–258.

Drahor, M. G.
2004 Application of the self-potential method to archaeological prospection: Some case histories. *Archaeological Prospection* 11(2): 77–105.

Drahor, M. G., A. L. Akyol, & N. Dilaver
1996 An application of the self-potential (SP) method in archaeogeophysical prospection. *Archaeological Prospection* 3(3): 141–158.

El-Gamili, M. M., A. S. El-Mahmoudi, S. S. Osman, A. G. Hassaneen, & M. A. Metwaly
1999 Geoelectric resistance scanning on parts of Abydos Cemetery Region, Sohag Governorate, Upper Egypt. *Archaeological Prospection* 6(4): 225–239.

Fox, R. C., G. W. Hohmann, T. J. Killpack, & L. Rijo
1980 Topographic effects in resistivity and induced-polarization surveys. *Geophysics* 45(1): 75–93.

Gaffney, C., & J. Gater
2003 *Revealing the buried past: Geophysics for archaeologists.* Strout: Tempus Publishing Ltd.

Griffiths, D. H., & R. D. Barker

1994 Electrical imaging in archaeology. *Journal of Archaeological Sciences* 21(2): 153–158.

Haigh, J. G. B.

1992 Automatic grid balancing in geophysical survey. In G. Lock and J. Moffett (eds.), *Computer applications and quantitative methods in archaeology 1991*, 191–196. BAR International Series 577. Oxford: Tempus Reparatum.

Harding, J.

2012 Henges, rivers and exchange in Neolithic Yorkshire. In A. Jones, J. Pollard, M. Allen, and J. Gardiner (eds.), *Image, memory and monumentality: Archaeological engagements with the material world (a celebration of the academic achievements of Professor Richard Bradley)*, 43–51. Oxford/Oakville: Oxbow Books and Prehistoric Society.

Hesse, A.

1966 The importance of climatic observations in archaeological prospecting. *Prospezione Archaeologische* 1: 11–13.

2000 Count Robert du Mesnil du Buisson (1895–1986), a french precursor in geophysical survey for archaeology. *Archaeological Prospection* 7(1): 43–49.

Hesse, A., A. Jolivet, & A. Tabbagh

1986 New prospects in shallow depth electrical surveying for archaeological and pedological applications. *Geophysics* 51(3): 585–594.

Jackson, J. D.

1975 *Classical electrodynamics*. New York: John Wiley and Sons.

Jordan, D.

2009 How effective is geophysical survey? A regional review. *Archaeological Prospection* 16(2): 77–90.

Kampke, A.

1999 Focused imaging of electrical resistivity data in archaeological prospecting. *Journal of Applied Geophysics* 41(2–3): 215–227.

Kaufman, A. A., & B. I. Anderson

2010 *Principles of electric methods in surface and borehole geophysics*. Amsterdam/London: Elsevier.

Keller, G., & F. C. Frischknecht
1966 *Electrical methods in geophysical prospecting* 10. Oxford: Pergamon Press Ltd.

Kvamme, K. L.
2006 Integrating multidimensional geophysical data. *Archaeological Prospection* 13(1): 57–72.

Leckebusch, J.
1998 Automatisierung von Radar- und Wiederstandsmessungen in Verbindung mit dreidimensionaler Radardatenverarbeitung. In *Unsichtbares sichtbar machen: Geophysikalische Prospektionsmethoden in der Archäologie*, 77–80. Stuttgart: Konrad Theiss Verlag.

Li, X., & H.-J. Götze
1999 Comparison of some gridding methods. *The Leading Edge* 18(8): 898–900.

Loke, M. H., I. Acworth, & T. Dahlin
2003 A comparison of smooth and blocky inversion methods in 2D electrical imaging surveys. *Exploration Geophysics* 34(3): 182–187.

Loke, M. H., & R. D. Barker
1995a Improvements to the Zohdy method for the inversion of resistivity sounding and pseudosection data. *Computers & Geosciences* 21(2): 321–332.
1995b Least-squares deconvolution of apparent resistivity pseudosections. *Geophysics* 60(6): 1682–1690.
1996 Rapid least-squares inversion of apparent resistivity pseudosections by a quasi-Newton method. *Geophysical Prospecting* 44(1): 131–152.

Luck, E., R. Gebbers, J. Ruehlmann, & U. Spangenberg
2009 Electrical conductivity mapping for precision farming. *Near Surface Geophysics* 7(1): 15–25.

Lynam, J. T.
1970 *Techniques of geophysical prospection as applied to near surface structure determination*. Thesis. Bradford: Department of Physics, University of Bradford.

Mathieson, I., E. Bettles, J. Dittmer, & C. Reader
1999 The National-Museums-of-Scotland Saqqara survey project, earth sciences 1990–1998. *Journal of Egyptian Archaeology* 85: 21.

Michot, D., Y. Benderitter, A. Dorigny, B. Nicoullaud, D. King, & A. Tabbagh

2003 Spatial and temporal monitoring of soil water content with an irrigated corn crop cover using surface electrical resistivity tomography. *Water Resources Research* 39(5): 1138.

Noel, M., & B. Xu

1991 Archaeological investigation by electrical resistivity tomography: A preliminary study. *Geophysical Journal International* 107(1): 95–102.

Novak, V.

2012 *Evapotranspiration in the soil-plant-atmosphere system*. New York: Springer.

Palacky, G. J.

1987 Resistivity characteristics of geologic targets. In M. N. Nabighian (ed.), *Electromagnetic methods in applied geophysics theory*, 53–129. Tulsa, OK: Society of Exploration Geophysicists.

Papadopoulos, N. G., P. Tsourlos, G. N. Tsokas, & A. Sarris

2006 Two-dimensional and three-dimensional resistivity imaging in archaeological site investigation. *Archaeological Prospection* 13(3): 163–181.

Parkyn, A.

2010 A survey in the park: Methodological and practical problems associated with geophysical investigation in a late Victorian municipal park. *Archaeological Prospection* 17(3): 161–174.

Richmond, I. A.

1949–1950 Excavations at the Roman fort of Newstead, 1947. *Proceedings of the Society of Antiquarians in Scotland* 84: 1–38.

Roy, A., & A. Apparao

1971 Depth of investigation in direct current methods. *Geophysics* 36(5): 943–959.

Schmidt, A.

2001 Visualisation of multi-source archaeological geophysics data. In M. Cucarzi and P. Conti (eds.), *Filtering, optimisation and modelling of geophysical data in archaeological prospecting*, 149–160. Rome: Fondazione Ing. Carlo M. Lerici.

2003 Remote sensing and geophysical prospection. *Internet Archaeology* 15. http://intarch.ac.uk/journal/issue15/schmidt_index.html.

Schmidt, A., R. A. E. Coningham, K. M. Strickland, & J. E. Shoebridge
2011 A pilot geophysical evaluation of the site of Tilaurakot, Nepal. *Ancient Nepal* 177: 1–16.

Schmidt, A., & E. Ernenwein
2011 Guide to good practice: Geophysical data in archaeology. *Archaeology Data Service / Digital Antiquity Guides to Good Practice.* http://guides.archaeologydataservice.ac.uk/g2gp/Geophysics_Toc.

Schmidt, A., T. Sutherland, & S. Dockrill
2006 Inside the mound: Geophysical surveys of the Scatness Iron-age broch, Shetland. In R. E. Jones and L. Sharpe (eds.), *Going over old ground*, 225–230. BAR British Series 416. Oxford: Archaeopress.

Scollar, I., A. Tabbagh, A. Hesse, & I. Herzog
1990 *Archaeological prospecting and remote sensing (Topics in remote sensing 2).* Cambridge: Cambridge University Press.

Somers, L. E., M. L. Hargrave, & J. E. Simms
2003 *Geophysical surveys in archaeology: Guidance for surveyors and sponsors.* ERDC/CERL SR-03-21. Arlington, VA: US Army Corps of Engineers. Engineer Research and Development Center.

Stoertz, C.
1997 *Ancient landscapes of the Yorkshire Wolds: Aerial photographic transcription and analysis.* Swindon: Royal Commission on Historical Monuments (England).

Sutherland, T. L., A. Schmidt, & S. Dockrill
1998 Resistivity pseudosections and their topographic correction: A report on a case study at Scatness, Shetland. *Archaeological Prospection* 5(4): 229–237.

Szymanski, J. E., & P. Tsourlos
1993 The resistive tomography technique for archaeology: An introduction and review. *Archaeologia Polona* 31: 5–32.

Tsokas, G. N., P. I. Tsourlos, & J. E. Szymanski
1997 Square array resistivity anomalies and inhomogeneity ratio calculated by finite-element method. *Geophysics* 62(2): 426–435.

Tsourlos, P., J. E. Szymanski, & G. N. Tsokas
2005 A generalized iterative back-projection algorithm for 2-D reconstruction of resistivity data: Application to data-sets from archaeological sites. *Journal of the Balkan Geophysical Society* 8(2): 37–52.

Tsourlos, P. I., & G. N. Tsokas
2011 Non-destructive Electrical Resistivity Tomography survey at the South Walls of the Acropolis of Athens. *Archaeological Prospection* 18(3): 173–186.

Werban, U., K. Kuka, & I. Merbach
2009 Correlation of electrical resistivity, electrical conductivity and soil parameters at a long-term fertilization experiment. *Near Surface Geophysics* 7(1): 5–14.

Witten, A. J.
2006 *Handbook of geophysics and archaeology (Equinox handbooks in anthropological archaeology)*. London/Oakville, CT: Equinox Pub.

Wynn, J. C.
1986 A review of geophysical methods used in archaeology. *Geoarchaeology* 1(3): 245–257.

Wynn, J. C., & S. I. Sherwood
1984 The Self-Potential (SP) Method: An inexpensive reconnaissance and archaeological mapping tool. *Journal of Field Archaeology* 11(2): 195–204.

Yilmaz, S., & N. Coskun
2011 A study of the terrain-correction technique for the inhomogeneous case of resistivity surveys. *Scientific Research and Essays* 6(24): 5213–5223.

Zohdy, A. A. R.
1989 A new method for the automatic interpretation of Schlumberger and Wenner sounding curves. *Geophysics* 54(2): 245–253.

INDEX

ABOUT THE AUTHOR

Armin Schmidt, honorary research fellow at the University of Bradford and Durham University, United Kingdom, is an archaeological geophysicist. He has applied novel methods of geophysical prospection worldwide, working as a researcher and United Nations Educational, Scientific, and Cultural Organization (UNESCO) consultant. He is the founder of the International Society for Archaeological Prospection (ISAP) and the Bradford Centre for Archaeological Prospection (B-CAP) and a cofounder of the Archaeology Data Service (ADS). He coauthored *Magnetometry for Archaeologists* (2008).

Lightning Source UK Ltd.
Milton Keynes UK
UKOW04n2042010415

248960UK00005B/99/P